ACE YOUR MIDTERMS & FINALS

INTRODUCTION TO BIOLOGY

Other books in the Ace Your Midterms and Finals Series include:

Ace Your Midterms and Finals: Introduction to Psychology

Ace Your Midterms and Finals: U.S. History

Ace Your Midterms and Finals: Fundamentals of Mathematics

Ace Your Midterms and Finals: Introduction to Physics

Ace Your Midterms and Finals: Principles of Economics

ACE YOUR MIDTERMS & FINALS

INTRODUCTION TO BIOLOGY

ALAN AXELROD, PH.D.

McGraw-Hill

New York San Francisco Washington, D.C. Auckland Bogotá
Caracas Lisbon London Madrid Mexico City Milan
Montreal New Delhi San Juan Singapore
Sydney Tokyo Toronto

Library of Congress Catalog Card Number: 99-070503

McGraw-Hill

A Division of The ***McGraw-Hill*** *Companies*

1 2 3 4 5 6 7 8 9 0 DOC/DOC 9 0 9 8 7 6 5 4 3 2 1 0 9

ISBN 0-07-007009-1

The sponsoring editor for this book was Barbara Gilson, the editing supervisor was Maureen Walker, and the production supervisor was Modestine Cameron. It was set in Minion by Carol Norton for The Ian Samuel Group, Inc.

Printed and bound by R. R. Donnelley & Sons Company.

This book is printed on recycled, acid-free paper containing a minimum of 50% recycled, de-inked fiber.

CONTENTS

HOW TO USE THIS BOOK

YOU KNOW THE DRILL. FIRST DAY IN A SURVEY, INTRODUCTORY, OR "CORE" COURSE: the professor talks about grading and, saying something about the value of the course in a program of "liberal education," declares that what she wants from her students is original thought and creativity and, above all, does not "teach for" the midterm and final.

Nevertheless, the course certainly *includes* one or two midterms and a final, which account for a very large part of the course grade. Maybe the professor can disclaim with a straight face *teaching for* these exams, but few students would deny *learning for* them.

True, you know that the purpose of an introductory course is to gain a useful familiarity with a certain field and not just to prepare for and do well on a couple or three exams. Yet the exams *are* a big part of the course, and, whatever you learn or fail to learn in the course, your performance as a whole is judged in large measure by your performance on these exams.

So the cold truth is this: More than anything else, curriculum core courses *are* focused on the midterm and final exams.

Now, traditional study guides are outlines that attempt a bird's-eye view of a given course. But *Ace Your Midterms and Finals: Introduction to Biology* breaks with this tradition by viewing course content through the magnifying lens of ultimate accountability: the course exams. The heart and soul of this book consists of eleven midterms and eleven finals prepared by *real* instructors, teaching assistants, and professors for *real* students in *real* schools.

Where did we get these exams? Straight from the professors and instructors themselves.

- All exams are real and have been used in real courses.
- All exams include critical "how-to" tips and advice from the creators and graders of the exams.
- All exams include actual answers.

Let's talk about those answers for a minute. In most cases, the answers are actual student responses to the exam. In some cases, however, the instructors and professors have created "model" or "ideal" answers. Usually, the answers included are A-level responses. Sometimes, however, they are not perfect (because they are real). In all cases, you'll find full commentary by the instructors, who point out what works (and why) and what could use improvement (and why—as well as how to improve it).

This book also contains more than the exams themselves.

- In "Part One: Preparing Yourself," you'll find how-to guidance on what biology professors look for, how to think like a biologist, how to study more effectively, and how to gain the performance edge when you take an exam.
- "Part Two: Study Guide" presents a quick-and-easy overview of the content of typical surveys of biology. It clues you in on what to expect in these courses.
- "Part Three: Midterms and Finals," is the exams themselves, grouped by college or university.
- In "Part Four: For Your Reference," you'll find a handy glossary of key terms in biology and a brief list of recommended reading.

What This Book *Is Not*

Ace Your Midterms and Finals: Introduction to Biology offers a lot of help to see you through to success in this important course or course sequence. But (as you'll discover when you read Part One) the book *cannot* take the place of

- Doing the assigned reading
- Keeping up with your work and study
- Attending class
- Taking good lecture notes
- Thinking about and discussing the topics and issues raised in class and in your books

Ace Your Midterms and Finals: Introduction to Biology is not a substitute for the course itself!

What This Book *Is*

Look, it's both cynical and silly to invest your time, brainpower, and money in a college course just so that you can ace a couple of exams. If you get A's on the midterm and final, but come away from the course having learned nothing, you've failed.

We don't want you to be cynical or silly. The purpose of introductory, survey, or "core" courses is to give you a panoramic view of the knowledge landscape of a particular field. The primary goal of the college experience is to acquire more than tunnel intelligence. It is to enable you to approach whatever field or profession or work you decide to specialize in from the richest, broadest perspective possible. College is education, not just vocational training.

We don't want you to "study for the exam." The idea is to study for "the rest of your life." You are buying knowledge with your time, your brains, and your money. It's an expensive and valuable commodity. Don't leave it behind you in the classroom at the end of the semester. Take it with you.

But even the starriest-eyed idealist can't deny that midterms and finals are a big part of intro courses and that even if your ambitions lie well beyond these exams (which they should!), performing well on them is necessary to realize those loftier ambitions.

Don't, however, think of midterms and finals as hurdles—obstacles—you must clear in order to realize your ambitions and attain your goals. The exams are there. They're real. They're facts of college life. You might as well make the most of them.

Use the exams to help you focus your study more effectively. Most people make the mistake of confusing *goals* with *objectives.* Goals are the big targets, the ultimate prizes in life. Objectives are the smaller, intermediate steps that have to be taken to reach those goals.

Success on midterms and finals is an objective. It is an important, sometimes intimidating, but really quite *doable* step toward achieving your goals. Studying for—working toward—the midterm or final is not a bad thing, as long as you keep in mind the difference between objectives and goals. In fact, fixing your eye on the upcoming exam will help you to study more effectively. It gives you a more urgent purpose. It also gives you something specific to set your sights on.

And this book will help you study for your exams more effectively. By letting you see how knowledge may be applied—immediately and directly—to exams, it will help you acquire that knowledge more quickly, thoroughly, and certainly. Studying these exams will help you to focus your study in order to achieve success on the exams—that is, to help you attain the objectives that build toward your goals.

—Alan Axelrod

CONTRIBUTORS

Richard Borowsky, *Associate Professor of Biology, New York University*

Carol Ann Kearns, *Associate Director, Environmental Residential Academic Program and Senior Instructor of Environmental, Population, and Organismic Biology, University of Colorado*

Michael Kutilek, *Assistant Professor of Biological Sciences, San Jose State University*

Elizabeth McGee, *Assistant Professor of Biological Sciences, San Jose State University*

Leathem Mehaffey, III, *Associate Professor of Biology, Vassar College*

Travis Phillip Multhaupt, *Graduate Teaching Assistant, Michigan State University*

Peggy Shadduck Palombi, *Assistant Professor of Biology, Transylvania University*

Javier Peñalosa, *Associate Professor of Biology, Buffalo State College*

Laurel Roberts, *Lecturer, Department of Biological Sciences, University of Pittsburgh*

Peter Sherman, *Assistant Professor of Biology, University of Louisville*

Janice Voltzow, *Associate Professor of Biology, University of Scranton*

ABOUT THE AUTHOR

Alan Axelrod, Ph.D., is the author of many books, including *Booklist* Editor's Choice *Art of the Golden West, The Penguin Dictionary of American Folklore,* and *The Macmillan Dictionary of Military Biography*. He lives in Atlanta, Georgia.

PART ONE

PREPARING YOURSELF

CHAPTER 1

INTRODUCTION TO GENERAL BIOLOGY: WHAT THE PROFESSORS LOOK FOR

BIOLOGY IS THE STUDY OF LIFE AND LIVING THINGS. WHAT COULD BE MORE RELEVANT or more interesting?

Still, it is possible to be disappointed or unpleasantly surprised if—

- You expect your introductory biology course to be a collection of nature films.
- You expect biology to be a kind of "nature appreciation course."
- You are unwilling to exercise your powers of observation.
- You are unwilling to approach with an open mind such subjects as sex, reproduction, and the origin and evolution of life.
- You are unwilling to deal with a host of new concepts and new words (many of them in scientific Latin).
- You are unwilling to exercise your memory.
- You don't like dealing with living things on a nitty-gritty level.
- You are not a good visual learner. Biology involves a lot of diagrams and drawings.
- You have taken biology as an alternative to physics or chemistry because you don't like mathematical equations and chemical formulas. As the science of life and living things, biology draws on many fields, including physics and especially chemistry. You will indeed encounter chemical formulas, and you will be asked to

The main course objective [of Biology 100, Principles of Biology] is to give students an appreciation of the principles of biology as they apply at the cellular and molecular level. All of the students are nonmajors. I want students to make connections and understand processes. This is more important than memorizing all the steps in a process—for example, knowing the Krebs cycle cold. Since I have to assume that this is my students' last (as well as their first) biology course, I don't worry too much about what they will "need to know" for further work in biology. Thus I hope the exams reward understanding rather than memorization.

—Javier Peñalosa, Associate Professor, Buffalo State College understand some chemical reactions fundamental to life. You will also learn about the basic structure and behavior of atoms and subatomic particles, as well as how atoms bond into molecules. Depending on the focus of the course you take, biochemistry—especially chemistry on a cellular level—may be a major part of the syllabus.

Regarding the two-semester General Biology sequence at the University of Colorado, the primary objective is to prepare students with a broad background of basic biological concepts. Students taking more advanced courses in biology refer back to their general biology course work repeatedly as they begin to delve more deeply into specific topics. Our department offers another biology sequence called "Biology, a Human Approach," for nonmajors who generally don't plan to take additional biology courses.

—Carol Ann Kearns,
Associate Director, Environmental
Residential Academic Program and
Senior Instructor of Environmental,
Population, and Organismic Biology,
University of Colorado

A Variety of Approaches

Let's talk about the vastness of biology as a field. In fact, biology is so vast that some universities have abandoned the idea of *a* Biology Department and, instead, have separate Botany, Zoology, and even Molecular Biology departments. Similarly, many introductory-level biology courses do indeed survey all the disciplines that biology comprises, often in a two-semester sequence. (Chapter 6 presents an overview of the major branches of biology.) But many other courses introduce biology by focusing on some more narrow aspect of life science—human biology, perhaps, or the cell, or ecology, or the diversity of organisms, or some other approach.

The point is that there is no universally accepted version of intro to biology. In fact, you may find that the biology department at your school offers more than one course to satisfy a survey or core requirement. You should:

- Read the course catalogue description(s) carefully.
- Talk to the instructor before you sign up.
- Find out what kind of science background (if any) is required, expected, or considered helpful.
- Decide how you feel about lab work. (Some folks are pretty squeamish.) Find out what role (if any) lab plays in the course.
- If multiple courses are offered, work with an advisor or instructor to choose the approach that is right for you.

In Bio 150 and 160, Foundations of Biology I and II, I expect students to master a body of information on cell structure, genetics (especially molecular), and evolution. Students should be able to organize material well enough to relate it to analogies and to make real-world applications. Students should be able correctly to answer compare-and-contrast questions.

—Laurel Roberts, Lecturer,
University of Pittsburgh

Great Expectations

Just as course offerings, focus, and approach vary, so do the instructor's expectations. Some expect students, even nonscience majors, to come to the course reasonably well prepared in science. Others have no such expectations.

Some instructors approach the survey course as a body of information to be absorbed. This is especially true in lecture-oriented courses. Others put more emphasis on lab work, outside reading, and independent research. Although some instructors want you to memorize and master extensive amounts of prescribed materials, others are more interested in your learning basic principles, which you are then expected to apply to practical problems. It is the case, however, that whatever else they look for, most scientists expect you to rely on your own initiative. In any science class, it helps to be a self-starter rather than someone who expects to be spoonfed the course contents.

In addition to attending class, taking lecture notes, reading the textbook, and doing all assigned exercises, your instructor expects you to

- ◆ think problems through
- ◆ pay close attention to lab work
- ◆ study any drawings and diagrams as carefully as you study text
- ◆ ask questions

Although, as mentioned, instructor expectations concerning your science background vary, at minimum you should have an understanding of the scientific method and its objectives and goals (see Chapter 6). You should also work to develop your powers of observation and your willingness to attend to detail. Moreover, you need a willingness to entertain new ideas and, while absorbing the details, you need to look beyond them to the "big picture." Biology is a science of systems—parts functioning within wholes. Get into the habit of learning how units function and work together in complete systems.

The best way to succeed on the exams and in the course is to put pen to paper. Take good notes in class and redraw—from memory—all relevant figures found in the textbook. If you own a tape recorder, do not take it to class! Give the tape recorder to your worst enemy.

Study lecture notes and read the book within 24 hours of attending the lecture. If you start studying for the exam two weeks in advance, you are probably already four weeks too late.

—Richard Borowsky, Associate Professor, New York University

I expect my students to appreciate the discipline of science in general, and how it impacts their lives on an everyday basis. This way, if students decide that science is not what they want to do for the rest of their lives, hopefully they will be educated enough to realize the importance of science and why it is necessary for the advancement of the human race.

—Travis Phillip Multhaupt, Graduate Teaching Assistant, Michigan State University

Many students claim to be "visual learners," but they tend to ignore illustrations.

—Javier Peñalosa, Associate Professor, Buffalo State College

Memorization and Study

Even for students who are fascinated by the life sciences, biology can present some formidable challenges. Every aspect of the science involves learning new terms—most of them of a technical nature, many of them of Latin derivation (difficult to pronounce, spell, and memorize), and many of them similar enough to cause confusion.

- ◆ Commit yourself to memorizing plenty of these terms.
- ◆ Wherever possible, associate a term with a formula or a picture. Don't try to learn a list of terms in a vacuum.
- ◆ Similarly, it is best to learn associated terms together. For example, the time to learn the meanings of *dendrite*, *axon*, *myelin sheath*, and *synapse* is when you are studying nerve cells and are learning how these structures work together as a system.

- Identify and pay special attention to the *key* terms. Your instructor will point out many of these. Obviously, too, if a term is used often in lecture or in your textbook, you must regard it as a key term, and you should become comfortable with it.
- The best way to learn strange new terms is to make them familiar by actually using them. Practice.

The study habits that work for other college courses work well for biology, too. Because each lecture introduces new concepts—and a set of new terms—it is very important to:

- Read textbook assignments *before* lectures.
- Keep up with your reading.
- Attend the lectures.
- Take good notes.

Primary course objectives of Biology 21, Human Biology, are to provide students with a general overview of key issues in human biology; to increase science literacy; and to encourage life-long learning of science. I want my students to internalize information acquired in class in order to make informed decisions about personal health issues and environmental issues, and to be informed voting citizens.

—Elizabeth McGee, Assistant Professor, San Jose State University

The main objectives of Biology 141 and 142, General Biology I and II, are to introduce students to some of the fundamental concepts of biology; to help them learn to read, think, and write critically; and to help them learn how to learn independently. I expect students to obtain a basic knowledge of some of the major concepts of biology, to begin to relate what happens at one level of organization to another (for example, to relate cellular respiration to the need for gas exchange systems), to think critically about what they read in a textbook or outside reading, and to begin to learn to organize for themselves the large amount of information they are exposed to over the course of two semesters.

—Janice Voltzow, Associate Professor, University of Scranton

Visual Learning

Remember that the diagrams you will encounter in your textbook and in lecture presentations are more than decorative illustrations. They are important tools for understanding major concepts. A good way to study diagrams is to try drawing from memory at least the major features of key diagrams. For example, you should be able to make a quick schematic sketch of a typical animal cell, with the major parts correctly labeled.

A Respect for Life

You can count on your biology instructor having one more expectation of you. Biology is a science, but, as a *life* science, it has a strong spiritual dimension. Most biology instructors are scientists who enter the field with a profound respect—even reverence—for life. Cultivating such a respect will not only enhance your appreciation and enjoyment of the course, it will put you more closely in tune with the instructor's attitude. And that is never a bad thing.

CHAPTER 2

KEYS TO SUCCESSFUL STUDY

THIS CHAPTER OUTLINES SKILLS THAT ARE INDISPENSABLE TO SUCCESSFUL STUDY, with special emphasis on skills important to the study of biology.

But let's not start thinking about biology just yet. Let's just start thinking. After all, isn't that what college and college courses are all about?

Well, not quite. *Think* about it.

For the ancient Greeks of Plato's day, about 428 to 348 B.C., "higher education" really was all about thinking. Through dialogue, back and forth, the teacher and his student *thought* about biology, mathematics, physics, the nature of reality—whatever. Perhaps the teacher evaluated the quality of his student's thought, but there is no evidence that Plato graded exams, let alone assigned the student a final grade for the course.

B.C. was a long time ago. Times have changed.

"Don't Study for the Grade!"

Today, you get graded. All the time, and on everything you do.

Now, most of the professors, instructors, and teaching assistants from whom you take your courses will tell you that the "real value" of the course is in its contribution to your "liberal education." A professor may even solemnly protest that she does *not* "teach for" the midterm and final. Nevertheless, the introductory, survey, and core courses almost always *include* at least one midterm and a final, and these almost always account for a very large part of the course grade. Even if the professor can disclaim with a straight face *teaching for* these exams, few students would deny *learning for* them.

The truth is this: More than anything else, most curriculum core courses are focused on the midterm and final exams.

"Grades Aren't Important"

Let's keep thinking.

You and, most likely, your family are investing a great deal of time and cash and sweat in your college education. It *would* be pretty silly if the payoff of all those resources were a letter grade and a numerical GPA. Ultimately, of course, the payoff is knowledge, a feeling of achievement, an intellectually and spiritually enriched life—*and* preparation for a satisfying and (you probably hope) financially rewarding career.

But the fact is that if you don't perform well on midterms and finals, your path to all these forms of enrichment will be blocked. And the fact *also* is that your performance is measured by grades. Sure, almost any reason you can think of for investing in college is more important than amassing a collection of A's and B's, but those stupid little letters are part of what it takes to get you to those other, far more important, goals.

"Don't Study *for* the Exam"

Most professors hate exams and hate grades. They believe that prodding students to pass tests and then evaluating their performance with a number or letter makes the whole process of education seem pretty trivial. Those professors who tell you that they "don't teach for the exam" may also advise you not to "study for the exam." That's not exactly what they mean. They want you to study, but to study in order to learn, not *for* the exam.

It's well-meaning advice, and it's true that if you study *for* the exam, intending to ace it, then promptly forget everything you've "learned," you are making a pretty bad mistake. Yet those same professors are part of a system that demands exams and grades, and if you don't study *for* the exam, the chances are very good that you won't make the grade and you won't achieve the "higher goals" you, your family, and *your professors* want for you.

Lose-Lose or Win-Win? Your Choice

When it comes to studying, especially in your introductory-level courses, you have some choices to make. You can decide grades are stupid, not study, and perform poorly on the exams. You can try not to study for the exams, but concentrate on higher goals, perform poorly on the exams, and never have the opportunity to reach those higher goals. You can study *for* the exams, ace them all, then flush the information from your memory banks, collect your A or B for the course, and move on without having learned a thing. These are all lose-lose scenarios, in which no one—neither you nor your teacher (nor your family, for that matter)—gets what he, she, or they really want.

Or you can go the win-win route.

We've used the words *goal* and *goals* several times. For an army general, winning the war is the goal, but to achieve this goal he must first accomplish certain *objectives,* such as winning battle number one, number two, number three, and so on. Objectives are intermediate goals or steps toward an ultimate goal.

Now, put exams in perspective. Performing well on an exam need not be an alternative to achieving higher goals, but should be an objective necessary to achieving those higher goals.

The win-win scenario goes like this: Use the fact of the exams as a way of focusing your study for the course. Focus on the exams as immediate objectives, crucial to achieving your ultimate goals. *Do* study for the exam, but not *for* the exam. *Don't* mistake the battle for the war, the objective for the goal; but *do* realize that you must attain the objectives in order to achieve the goal.

And *that,* ladies and gentlemen, is the purpose of this book:

- To help you ace the midterm and final exams in introductory-level biology courses . . .
- . . . without forgetting everything you learned after you've aced them.

This guide will help you use the exams to master the course material. This guide will help you make the grade—*and* actually learn something in the process.

Focus

How many times have you read a book, word for word, finished it, and closed the cover—only to realize that you've learned almost nothing from it? Unfortunately, it's something we all experience. It's not that the material is too difficult or that it's over our heads. It's that we mistake reading for studying.

For many of us, reading is a passive process. We scan page after page, the words go in, and, alas, the words seem to go out. The time, of course, goes by. We've read the book, but we've *retained* all too little.

Studying certainly involves reading, but reading and studying aren't one and the same activity. Or we might put it this way: studying is intensely focused reading.

How do you *focus* reading?

Begin by setting objectives. Now, saying that your objective in reading a certain number of chapters in a textbook or reading your lecture notes is to "learn the material" is not very useful. It is an obvious but vague *goal* rather than a well-defined objective.

Why not let the approaching exam determine your objective?

"I will read and retain the stuff in chapters ten through twenty because that's what's going to be on the exam, and I want to ace the exam."

Now you at least have an objective. Accomplish this *objective*, and you will be on your way to achieving the *goal* of "learning the material."

◆ An *objective* is an immediate target. A *goal* is for the long term.

Concentrate

To move from passive reading to active study requires, first of all, concentration. Setting up objectives (immediate targets) rather than looking toward goals (long-term targets) makes it much easier for you to concentrate. Few of us can (or would) put our personal lives "on hold" for four years of college, several years of graduate school, and X years in the working world in order to concentrate on achieving a *goal*. But just about anyone can discipline himself or herself to set aside distractions for the time it takes to achieve the *objective* of studying for an exam.

Step 1. Find a quiet place to work.

Step 2. Clear your mind. Push everything aside for the few hours each day that you spend studying.

Step 3. Don't daydream—*now*. Daydreaming, letting your imagination wander, is actually essential to real learning. But, right now, you have a specific objective to attain. This is not the time for daydreaming.

Step 4. Deal with your worries. Those pressing matters that you can do something about *right now*, take care of. Those you can't do anything about, push aside for now.

Step 5. Don't *worry* about the exam. Take the exam seriously, but don't fret. Instead of worrying about the prospect of failure, use your time to eliminate failure as an option.

Plan

Let's go back to that general who knows the difference between objectives and goals. Chances are he or she also knows that you'd better not march off to battle without a plan. Remember, you don't want to *read*. You want to *study*. This requires focusing your work with a plan.

The first item to plan is your time.

Step 1. Dedicate a notebook or organizer-type date book to the purpose of recording your scheduling information.

Step 2. Record the following:

a. Class times

b. Assignment due dates

c. Exam dates

d. Extracurricular commitments

e. Study time

Step 3. Inventory your various tasks. What do you have to do today, this week, this month, this semester?

TIP: If you are in doubt about what tasks to assign the highest priority, it is generally best to allot the most time to the most complex and difficult tasks and to get these done first.

Step 4. Prioritize your tasks. Everybody seems to be grading you. Now's *your* chance to grade the things they give you to do. Label high-priority tasks "A," middle-priority tasks "B," and lower-priority tasks "C." This will not only help you decide which things to do first, it will also aid you in deciding how much time to allot to each task.

Step 5. Enter your tasks in your scheduling notebook and assign order and duration to each according to its priority.

Step 6. Check off items as you complete them.

Step 7. Keep your scheduling book up to date. Reschedule whatever you do not complete.

Step 8. Don't be passive. Actively *monitor* your progress toward your objectives.

Step 9. Don't be passive. Arrange and rearrange your schedule to get the most time when you need it most.

Packing Your Time

Once you have found as much time as you can, pack it as tightly as you can.

Step 1. Assemble your study materials. Be sure you have all necessary textbooks and notes on hand. If you need access to library reference materials, study in the library. If you need access to reference materials on the Internet, make sure you're at a computer.

Step 2. Eliminate or reduce distractions.

Step 3. Become an efficient reader and note taker.

An Efficient Reader

Step 3 requires further discussion. Let's begin with the way you read.

Nothing has greater impact on the effectiveness of your studying than the speed and comprehension with which you read. If this statement prompts you to throw up your hands and wail, "I'm just not a fast reader," don't despair. You can learn to read faster and more efficiently.

Consider taking a "speed reading" course. Take one that your university offers or endorses. Most of the techniques taught in the major reading programs actually do work. Alternatively, do it yourself.

TIP: Are you a—ugh!—*vocalizer*? A vocalizer is a reader who, during "silent" reading, either mouths the words or says them mentally. Vocalizing greatly slows reading, often reduces comprehension, and is just plain tiring. Work to overcome this habit—except when you are trying to memorize some specific piece of information. Many people do find it helpful to say over a sentence or two in order to memorize its content. Just bear in mind that this does not work for more than a sentence or paragraph of material.

Step 1. When you sit down to read, try consciously to force your eyes to move across the page faster than normal.

Step 2. Always keep your eyes moving. Don't linger on any word.

Step 3. Take in as many words at a time as possible. Most slow readers aren't slow-witted. They've just been taught to read word by word. Fast and efficient readers, in contrast, have learned to read by taking in groups of words. Practice taking in blocks of words.

Step 4. Build on your skills. Each day, push yourself a little harder. Move your eyes across the page faster. Take in more words with each glance.

Step 5. Resist the strong temptation to fall back into your old habits. Keep pushing.

When you review material, consider skimming rather than reading. Hit the high points, lingering at places that give you trouble and skipping over the stuff you already know cold.

TIP: The physical act of underlining actually helps you memorize material more effectively—though no one is quite sure why. Furthermore, underlining makes review skimming more efficient and effective.

An Interactive Reader

Early in this chapter we contrasted passive reading with active studying. A highly effective way to make the leap from passivity to activity is to become an *interactive* reader.

Step 1. Read with a pencil in your hand.

Step 2. Use your pencil to underline key concepts. Do this consistently. (That is, *always* read with a pencil in your hand.) Don't waste your time with a ruler; underscore freehand.

Step 3. Underline only the key concepts. If everything seems important to you, then up the ante and underline only the absolutely *most* important passages.

Step 4. If you prefer to highlight material with a transparent marker (a Hi-Liter, for example), fine. But you'll still need a pencil or pen nearby. Carry on a dialog with your books by writing condensed notes in the margin.

Step 5. Put difficult concepts into your own words—right in the margin of the book. This is a great aid to understanding and memorization.

TIP: At least one contributor to this book advises against using Hi-Liter-style markers to underscore books and notes, because doing so may discourage you from writing notes in the margin. If holding a Hi-Liter means that you won't also pick up a pencil to engage in a lively dialogue with books or notes, then it is best to lay aside the highlighter and take the more *actively* interactive route.

Step 6. Link one concept to another. If you read something that makes you think of something else related to it, make a note. The connection is almost certainly a valuable one.

Step 7. Comment on what you read.

Taking Notes

The techniques of underlining, highlighting, paraphrasing, linking, and commenting on textbook material can also be applied to your classroom and lecture notes.

Of course, this assumes that you have taken notes. There are some students who claim that it is easier for them to listen to a lecture if they *avoid* taking notes. For a small minority, this may be true; but the vast majority of students find that note taking is essential when it comes time to study for midterms and finals. This does not mean that you should be a stenographer or court reporter, taking down each and every word. To the extent that it is possible for you to do so, absorb the lecture *in your mind*, then jot down major points, preferably in loose outline form.

TIP: Some students are reluctant to write in their textbooks, because it reduces resale value. True enough. But is it worth an extra five dollars at the end of the semester if you don't get the most out of your multi-thousand-dollar and multi-hundred-hour investment in the course?

Become sensitive to the major points of the lecture. Some lecturers will come right out and tell you, "The following three points are key." That's your cue to write them down. Other cues include:

- **Repetition**. If the lecturer repeats a point, write it down.
- **Excitement.** If the lecturer's voice picks up, if his or her face becomes suddenly animated, if, in other words, it is apparent that the material is at this point of particular interest to the speaker, your pencil should be in motion.
- **Verbal cues.** In addition to such verbal elbows in the rib as "this is important," most lecturers underscore transitions from one topic to another with phrases such as "Moving on to . . ." or "Next, we will . . ." or the like. This is your signal to write a new heading.
- **Slowing down.** If the lecturer gives deliberate verbal weight to a word, phrase, or passage, make a note of it.
- **Visual aids.** If the lecturer writes something on a blackboard or overhead projector or in a computer-generated presentation, make a note.

Filtering Notes

Some students take neat notes in outline form. Others take sprawling, scrawling notes that are almost impossible to read. Most students can profit from *filtering* the notes they take. Usually, this does *not* mean rewriting or retyping your notes. Many instructors agree that this is a waste of time. Instead, they advise, underscore the most important points, filtering out the excess.

If you have taken notes on a notebook or laptop computer, consider arranging the notes in clear outline form. If you have handwritten notes, however, it may not be worth the time it takes to create a neat outline. Instead, spend that time merely underlining or highlighting the most important concepts.

TIP: Note taking is especially important in the sciences, because new discoveries are continually being made, many of which relate to fundamental issues relevant even in introductory courses. Lectures are up-to-date, whereas your textbook is obsolete the moment it is printed.

Build Your Memory

Just as a variety of speed-reading courses are available, so a number of memory-improvement courses, audio tapes, and books are on the market. It might be worth your while to scope some of these out, especially for a memory-intensive course like biology or if you are planning ultimately to go into a field that requires the memorization of a lot of facts. In the meantime, here are some suggestions for building your memory:

- Be aware that most so-called memory problems are really learning problems. You "forget" things you never *learned* in the first place. This is usually a result of passive reading or passive attendance at lectures—the familiar in-one-ear-and-out-the-other syndrome.
- Memorization is made easier by two seemingly opposite processes. First, we tend to remember information for which we have a context. It will indeed be hard to remember "a bunch of formulas and anatomical parts" if you try to study these out of context—as a list rather than as parts of an interrelated system of thought.
- Second, memorization is often made easier if we break down complex material into a set of key phrases or words. You may find it easier to memorize the key points relating to inflation and unemployment than a narrative description of these two phenomena.
- It follows, then, that the best way to build your memory where a certain subject is concerned is to try to understand information in context. Get the "big picture."
- It also follows that, even if you have the big picture, you may want to break down key concepts into a few key words or phrases.

TIP: No one can tell you just how much to write, but bear this in mind: Most lecturers read from notes rather than fully composed scripts. Four double-spaced typewritten or word-processed pages (about a thousand words *in note form*) represent sufficient note material for an hour-long lecture. Ideally, you might aim at producing about 75 percent of this word count in the notes you take—perhaps 750 words during an hour-long lecture.

TIP: While many instructors do not recommend rewriting or typing up your lecture notes, this may be useful in a subject like biology, which requires memorizing a good deal of technical information and specialized vocabulary. You may want to give this tactic a try. If it helps, use it. However, if you find that rewriting or typing your notes does not help you, drop the practice.

How We Forget

It is always better to keep up with class work and study than it is to fall behind and desperately struggle to catch up. This said, it is nevertheless true

TIP: Before you tote your laptop or notebook computer to class, make certain that the instructor approves of such note-taking devices in the classroom. Most lecturers have no problem with these, but some find the tap-tap-tapping of maybe more than a hundred students distracting.

And should you bring a tape recorder to class? The short answer is, probably not. To begin with, some instructors object to having their lectures recorded. Even more important, however, is the tendency to complacency a tape recorder creates. You might feel that you don't have to listen very carefully to the "live" lecture, since you're getting it all on tape. This is a mistake, since the live presentation is bound to make a greater impression on you, your mind, and your memory than a recorded replay.

that most forgetting occurs within the first few days of exposure to new material. That is, if you "learn" 100 facts about Subject A on December 1, you may forget 20 of those facts by December 5 and another 10 by December 10, but by March 1 you may still remember 50 facts. The curve of your forgetting tends to flatten. Eventually, you may forget all 100 facts, but you will forget fewer and fewer each week.

Now, what does this mean to you?

It means that you need to review material you learned earlier in the course. You cannot depend on having mastered it forever by having studied two, three, four, or more weeks earlier.

The Virtues of Cramming

Ask any college instructor about last-minute cramming for an exam, and you'll almost certainly get a knee-jerk condemnation of the practice. But maybe it's time to think beyond that knee jerk.

Let's get one thing absolutely straight. You cannot expect to pack a semester's worth of studying into a single all-nighter. It just isn't going to happen. However, cramming can be a valuable *supplement* to a semester of conscientious studying.

- You forget the most within the first few days of studying. (Or have you forgotten?)

Well, if you cram the night before the exam, those "first few days" won't fall between your studying and the exam, will they?

- Cramming creates a sense of urgency. It brings you face to face and toe to toe with your objective. Urgency concentrates the mind.
- Assuming you aren't totally exhausted, material you study within a few hours of going to bed at night is more readily retained than material studied earlier in the day.

Burning the midnight oil may not be such a bad idea.

Cramming Cautions

Then again, staying up late before a big exam may not be such a hot idea, either. Don't do it if you have an early-morning exam. And don't transform cramming into an all-nighter. You almost certainly need *some* sleep to perform competently on tomorrow's exam.

Remember, too, that while cramming creates a sense of urgency, which may stimulate and energize your study efforts, it may also create a feeling of panic, and panic is never helpful.

Cramming is *not* a substitute for diligent study throughout the semester. But just because you have studied diligently, don't shun cramming as a *supplement* to regular study, a valuable means of refreshing the mind and memory.

TIP: Memorization is important, make no mistake, especially in biology. However, it *is* usually overrated. Virtually all of the instructors and professors who have contributed to this book counsel students to *think* rather than merely memorize. When grading exams, most biology professors look for evidence of thought—of a thoughtful grasp of broad and basic concepts—rather than a regurgitation of facts memorized rote fashion.

Polly Want a Cracker?

We've been talking a lot about memory and memorization. It's an important subject and, for just about any course of study, an important skill. Some subjects—biology included—are more fact-and-memory intensive than others. However, beware of relying too much on simple, brute memory. Try to assess what the professor really wants: students who demonstrate on exams that they have "absorbed" the facts he or she and the textbooks have dished out, or students who demonstrate such skills as critical thinking, synthesis, analysis, imagination, and so on. Depending on the professor's personal style and the kind of exam he or she gives (predominantly essay vs. predominantly multiple choice, for example), you may actually be penalized for "parroting" lectures. ("I know what *I* think. I want to know what *you* think.")

TIP: Any full-time college student studies several subjects each semester. This makes you vulnerable to interference—the possibility that learning material from one subject will interfere with learning material in another subject. Interference is usually at its worst when you are studying two similar or related subjects. If possible, arrange your study time so that work on similar subjects is separated by work on an unrelated one.

Use *This* Book (and Get Old Exams)

One way to judge what the professor values and expects is to pay careful attention in class. Is discussion invited? Or does the course go by the book and by the lecture? Also valuable are previous exams. Many professors keep these on file and allow students to browse through them freely. Fraternities, sororities, and formal as well as informal study groups sometimes maintain such files, too. These days, previous exams may even be posted on the university department's World Wide Web site.

Of course, you are holding in your hand a book chockful of sample and model midterm and final exams. Read them. Study them. And let them focus your study and review of the course.

Study Groups: The Pro and The Con

In the old days (whenever that was), it was believed that teachers *taught* and students *learned*. More recently, educators have begun to wonder whether it is possible to *teach* at all. A student, they say, *learns* by teaching himself or herself. The so-called teacher (who might better be called a "learning facilitator") helps the student teach himself or herself.

TIP: If you *hate* cramming, don't do it. It's not for you, and it will probably only raise your anxiety level. Get some sleep instead.

Well, maybe this is all a matter of semantics. Is there really a difference between *teaching* and *facilitating learning*? And between *learning* and *teaching oneself*? The more important point is that the focus in education has turned away from the teacher to the student, and students, in turn, have often responded by organizing study groups, in which they help each other study and learn.

These can be very useful:

TIP: Many memory experts suggest that you try to put the key terms you identify into some sort of sentence; then memorize the sentence. Others suggest creating an acronym out of the initials of the key words or concepts. No one who lived through the Watergate scandal in the 1970s can forget that President Nixon's political campaign was run by CREEP (Committee to *RE*-Elect the President), and even nonscience majors remember one way that biologists traditionally classify living things by memorizing a sentence such as this: *Ken Put Candy On Fred's Green Sofa* (kingdom, phylum, class, order, family, genus, species).

- ◆ In the so-called real world (that is, the world after college), most problems are solved by teams rather than individuals.
- ◆ Many people come to an understanding of a subject through dialogue and question and answer.
- ◆ Studying in a group (or even with one partner) makes it possible to drill and quiz one another.
- ◆ In a group, complex subjects can be broken up and divided among members of the group. Each one becomes a "specialist" on some aspect of the subject, then shares his or her knowledge with the others.
- ◆ Studying in a group may improve concentration.
- ◆ Studying in a group may reduce anxiety.

Not that study groups are without their problems:

- ◆ All too often, study groups become social gatherings, full of distraction (however pleasant) rather than study. This is the greatest pitfall of a study group.
- ◆ All members of the group must be highly motivated to study. If not, the group will become a distraction rather than an aid to study, and it is also likely that friction will develop among the members, some of whom will feel burdened by "freeloaders."
- ◆ The members of the study group must not only be committed to study, but to one another. Study groups fall apart—bitterly—if members, out of a sense of competition, begin to withhold information from one another. This *must* be a Three Musketeers deal—all for one and one for all—or it is worse than useless.
- ◆ The group may promote excellence—or it may agree on mediocrity. If the latter occurs, the group will become destructive.

In summary, study groups tend to bring out the members' best as well as worst study habits. It takes individual and collective discipline to remain focused on the task at hand, to remain committed and helpful to one another, to insist that everyone shoulder his or her fair share, and to insist on excellence of achievement as the only acceptable standard—or, at least, the only valid reason for continuing the study group.

CHAPTER 3

SECRETS OF SUCCESSFUL TEST TAKING

SOMETIMES IT SEEMS THAT THE DIFFERENCE BETWEEN ACADEMIC SUCCESS AND something less than success is not smarts versus nonsmarts or even study versus nonstudy, but simply whether or not a person is "good at taking tests." That phrase—good at taking tests—was probably first heard back when the University of Bologna opened for business late in the eleventh century. The problem with phrases like this is that they are true enough, yet not very helpful.

Fact: Some people *are* and some people are *not* good at taking tests.

So what? Even if successful test taking doesn't come to you easily or naturally, you can improve your test-taking skills. Now, if you happen to have a knack for taking tests, well, congratulations! But that won't help you much if you neglect the kind of preparation discussed in the previous chapter.

Why Failure?

In analyzing performance on most tasks, it is generally better to begin by asking what you can do to succeed. But, in the case of taking tests, success is largely a question of *avoiding* failure. So let's begin there.

When the celebrated bank robber Willy "The Actor" Sutton was caught, a reporter asked him why he robbed banks. "Because that's where they keep the money," the handcuffed thief replied. At least one answer to the question of why some students perform poorly on exams is just as simple: "Because they don't know the answers."

There is no "magic bullet" in test taking. But the closest thing to one is *knowing the course material cold.* Pay attention, keep up with reading and other assignments, attend class, listen in class, take effective notes—in short, follow the recommendations of the previous chapter—and you will have taken the most important step toward exam success.

Yet have you ever gotten your graded exam back, with disappointing results, read it over, and found question after question you now realize you *could* have answered correctly?

"I *knew* that!" you exclaim, smacking yourself in the forehead.

What happened? You *really* did prepare. You *really* did know the material. What happened?

TIP: Most midterm and final exams really *are* representative of the course. If you have mastered the course material, you will almost certainly be prepared to perform well on the examination. Very few instructors purposely create deceptive exams or trick questions or even questions that require you to think beyond the course. Most instructors are interested in creating exams that help you and them evaluate your level of understanding of the course material. Think of the exams as natural, logical features of the course, not as sadistic assignments designed to trip you up. Remember, your success on an examination is also a measure of the instructor's success in presenting complex information. Very few teachers can—or want to—build careers on trying to *fail* their students.

Anxiety Good and Bad

The great American philosopher and psychologist William James (1842-1910) once advised his Harvard students that an "ounce of good nervous tone in an examination is worth many pounds of . . . study." By "good nervous tone" James meant something very like anxiety. You *should not* expect to feel relaxed just before or during an exam. Anxiety is natural.

Anxiety is "natural" because it is helpful. "Good nervous tone," alert senses, sharpened perception, adrenalin-fueled readiness for action are *natural* and *healthy* responses to demanding or threatening situations. We are animals, and these are reactions we share with other animals. The mongoose that relaxes when it confronts a cobra is a dead mongoose. The student who takes it easy during the bio final . . . Well, the point is neither to fight anxiety nor to fear it. Accept it, and even welcome it as an ally. Unlike our hominid ancestors of distant prehistory, we no longer need the biological equipment of anxiety to help us fight or fly from the snapping saber teeth of some animal of prey, but every day we do face challenges to our success. Midterms and finals are just such challenges, and the anxiety they provoke is real, natural, and unavoidable. It may even help us excel.

What good can anxiety do?

- Anxiety can focus our concentration. It can keep the mind from wandering. This makes thought easier, faster, and, often, more acute and effective.
- Anxiety can energize us. We've all heard stories about a 105-pound mother who is able to lift the wreckage of an automobile to free her trapped child. This isn't fantasy. It really happens. And just as adrenalin can provide the strength we need when we need it most, it can enhance our ability to think under pressure.

- Anxiety moves us along. We work faster than when we are relaxed. This is valuable since, in most midterms and finals, limited time is part of the test.
- Anxiety prompts us to take risks. We've all been in classes in which the instructor has a terrible time trying to get students to speak and discuss and venture an opinion. "Come on, come on," the poor prof protests, "this is like pulling teeth!" Yet, when exam time comes, all heads are bent over bluebooks or scantron answer sheets, and the answers—*some* kind of answers—are flowing forth or, at least, grinding out. Why? Because the anxiety of the exam situation overpowers the inertia that keeps most of us silent most of the time. We take the risks we have to take. We answer the questions.
- Anxiety can make us more creative. This is related to risk taking. "Necessity," the old saying goes, "is the mother of invention." Phrased another way: *we do what we have to do.* Under pressure, many students find themselves taking fresh and creative approaches to problems.

So don't shun anxiety. But, unfortunately, the scoop on anxiety isn't all good news, either.

Anxiety evolved as a mechanism of *physical* survival. Biologists and psychologists talk of the "fight or flight" response. Anxiety prepares a threatened animal either to fight the threat or to flee from it. The action is physical and, typically, very short term. In our "civilized" age, the threats are generally less physical than intellectual and emotional, and they tend to be of longer duration than a physical fight or a physical flight. This means that the anxiety mechanism does not always work to enhance our chances for "survival" or, at least, our chances to survive the course by performing well on exams. Some of us are better than others at adapting the *physical* benefits of anxiety to the *intellectual* and *emotional* challenges of an exam. Some of us, unfortunately, are unable to benefit from anxiety, and, for still others of us, anxiety is downright harmful. Here are some of the negative effects anxiety may have on exam performance:

- Anxiety can make it difficult to concentrate. True, anxiety focuses concentration. But if it focuses concentration on the anxious feelings themselves, you will have less focus left over for the exam. Similarly, anxiety may cause you to focus unduly on the perceived consequences of failure.
- Anxiety causes carelessness. If anxiety can prompt you to take creative risks, it can also cause you to rush through material and, therefore, to make careless mistakes or simply to fail to think through a problem or question.
- Anxiety distorts focus. Anxiety may impede your judgment, causing you to give disproportionate weight to relatively unimportant matters. For example,

you may become fixated on solving a lesser problem at the expense of a more important one. This is related to the next point.

- Anxiety may distort your perception of time. You may think you have more or less of it than you really do. The result may be too much time spent on a minor question at the expense of a major one.
- Anxiety tends to be cumulative. Many test takers have trouble with a question early in the exam, then devote the rest of the exam to worrying about it instead of concentrating on the *rest of the exam.*
- Anxiety drains energy. For short periods of time, anxiety can be energizing and invigorating. But if anxiety becomes chronic, it begins to tire you out. You do not perform as well.
- Anxiety can keep you from getting the rest you need. If it is generally unwise to stay up all night *studying* for an exam, how much less wise is it to stay up uselessly *worrying* about one?

TIP: Exam questions are battles in a war. No general expects to win every battle. Accept your losses and move on. Dwell on your losses, and you will continue to lose.

How can you combat anxiety?

Step 1: *Don't* fight it. Accept it. Remember, anxiety is a *natural* response to a stressful situation. Remember, too, that some degree of anxiety aids performance. Try to learn to accept anxiety and use it. Let it sharpen your wits and stoke the fires of your creativity.

Step 2: Don't worry about how you feel. Focus on the task. Usually, you will feel better once you overcome the initial jitters and inertia. William James, who lauded "good nervous tone," also once observed that we do not run because we are frightened, but we are frightened because we run. If you concentrate on your fear and act as if you are afraid, you will become even more fearful.

Step 3: Prepare for the exam. Do whatever you must do to master the material. Build confidence in your understanding of the course, and your anxiety should be reduced.

Step 4: Get a good night's sleep before the exam.

TIP: More serious stimulant drugs ("speed," "uppers") are *never* a good idea. They are both illegal and dangerous (possibly even deadly), and, for that matter, their effect on exam performance is unpredictable. The chances are they will impede performance rather than aid it (though you may erroneously *feel* that you are doing well). Also avoid over-the-counter stimulants. These are caffeine pills and will probably increase anxiety rather than improve performance.

Step 5: Avoid coffee and other stimulants. Caffeine tends to increase anxiety. (However, if you are a caffeine fiend, don't pick the day or two before a big exam to kick the habit. You *will* suffer withdrawal symptoms.)

Step 6: Try to get fresh air shortly before the exam. This is especially valuable if you have been cooped up for a long period of study. Take a walk. Get a look at the wider world for a few minutes.

TIP: Don't make the mistake of devoting all of your time to trying to make *last-minute* repairs to weak spots ("I've got one hour to read that textbook I should have been reading all along!") only to ignore your strengths. Develop your strengths. With any luck at all, the exam will give you an opportunity to show yourself at your best—not just trip you up at your worst. Be as prepared as you can be, but, remember, there is nothing wrong with excelling in a particular area. Play to your strengths, not your weaknesses.

Have a Plan, Make a Plan

A large component of destructive anxiety—probably the largest—is fear of the unknown. Reduce anxiety by taking steps to reduce the component of the unknown.

Step 1: To repeat—Do whatever is necessary to master the material on which you will be examined.

Step 2: Use the exams in this book to familiarize yourself with the kinds of exams you are likely to encounter.

Step 3: If possible, examine old exams actually given in the course.

Let's pause here before going on to Step 4. Just reading over the exams in this book or leafing through exams formerly given in the course will not help you much. Analyze:

- The types of questions asked. Are they essay or objective questions? (We'll discuss these shortly.) Do they call for "regurgitation" of memorized material, or are they more "think"-oriented, requiring significant initiative to answer?
- Don't just predict which questions you could and could not answer. Try actually answering some of the questions.
- If you are looking at sample exams with answers, evaluate the answers. How would you grade them? What would you do better?
- Don't just stand there, do something. If your analysis of the sample exams or old exams reveals areas in which you are weak, address those weaknesses.

An effective way to reduce the unknown is to create a plan for confronting it. Let's go on to Step 4.

Step 4: Make a plan. Begin *before* the exam. Decide what areas you need to study hardest. Based on your textbook notes and—especially—on your lecture notes, try to anticipate what kinds of questions will be asked. Work up answers or sketches of answers for these.

Step 5: Make sure you've done the simple things. The night before your exam, make certain that you have whatever equipment you'll need. If you will be allowed to use reference materials, bring them. If you are permitted to write on a laptop or notebook computer, make certain your batteries are fully charged. If you are writing the exam longhand, make certain you have pens, pencils, paper. Bring a watch.

Step 6: Expect a shock. The first sight of the exam usually packs a jolt. At first sight, questions may draw a blank from you. Questions you were sure would appear on the exam will be absent, and some you never expected will be staring you in the face. Don't panic. *Everybody feels this way.*

TIP: When you study for an exam, it is usually best to assign high priority to the most complex and difficult issues, devote ample time to these, and master them first. When you *take* an exam, however, and you are under time pressure, tackle first what you can most readily and thoroughly answer, then go on to more doubtful tasks. Your professor will be more favorably impressed by good or correct answers than by failed attempts to answer questions you find difficult.

Step 7: Write nothing yet. Read through the exam thoroughly. Be certain that you a) understand any instructions and b) understand the questions.

Step 8: If you are given a choice of which questions to answer, choose them now. Unless the questions vary in point value assigned to them, choose those that you feel most confident about answering. Don't challenge yourself.

Alternative Step 8: If you are required to answer all the questions, identify those about which you feel most confident. Answer these first.

Step 9: After you have surveyed the exam, create a "time budget": note—jot down—how much time you should give to each question.

Step 10: Reread the question before you begin to write. Then plan your answer.

Plan Your Answer

Perhaps you have heard a teacher or professor comment on the exam he or she has just handed out: "The answers are in the questions." This kind of remark is more helpful than it may at first seem.

Begin by looking for the key words in the question. These are the verbs that *tell* you what to do, and they typically include:

- Compare
- Contrast
- Criticize
- Define
- Describe
- Discuss
- Evaluate
- Explain
- Illustrate
- Interpret
- Justify
- Outline
- Relate
- Review

- State
- Summarize
- Trace

You will find most of these key words in essay questions rather than in short-answer, multiple-choice, or fill-in-the-blank questions, so we will have much more to say about the key words in Chapter 5, which is devoted to answering the essay exam. But be aware that the following key words are often found even in quite brief short-answer questions:

- ◆ To **define** something is to state the precise meaning of the word, phrase, or concept. Be succinct and clear.
- ◆ To **illustrate** is to provide a specific, concrete example.
- ◆ To **outline** is to provide the main features or general principles of a subject. This need not be in paragraph or essay form. Often, outline format is expected.
- ◆ To **state** is similar to **define**, though a statement may be even briefer and usually involves delivering up something that has been committed to memory.
- ◆ To **summarize** is briefly to state—in sentence form—the major points of an argument or position or concept, the principal features of an event, or the main events of a period.

As has just been mentioned, advice on answering essay examinations is the subject of Chapter 5. For the moment, just be aware that you will want to budget time for creating a scratch outline of your essay answer.

Approaching the Short-Answer Test

While the value of planning is more or less obvious in the case of essay exams, you will also find it valuable in "objective," multiple-choice, or other short-answer exams. These are of two major kinds:

1. **Recall exams** include questions that call for a single short answer (usually there is a single "correct" answer) and fill-in-the-blank questions, in which you are asked to supply missing information in a statement or sentence.
2. **Recognition exams** include multiple-choice tests, true-false tests, and matching tests (match one from column A with one from column B).

If the exam is a long one and time is short, invest a few minutes in surveying the questions, so that you can be certain to answer those you are confident of, even if they come near the end of the exam.

Be prepared to answer multiple-choice questions through a process of elimination, if necessary. Usually, even if you are uncertain of the one correct answer among a choice of five, you *will* be able to eliminate one, two, or three answers you know are *incorrect.* This at least increases your odds of giving a correct response.

Unless your instructor has specifically informed you that he or she is penalizing guesses (actually taking points away for incorrect responses versus awarding zero points to unanswered questions), do guess the answers even to those questions that leave you in the dark.

Plan your responses to true-false questions carefully. Look for telltale qualifying words, such as *all, always, never, no, none,* or *in all cases.* Questions with such absolute qualifiers *often* require an answer of *false,* since relatively few general statements are always either true or false. Conversely, questions containing such qualifiers as *sometimes, usually, often,* and the like are frequently answered correctly with a response of *true.*

A final word on guessing: *First guess, best guess.* Statistical evidence consistently shows that a first guess is more likely to be right than a later one. Obviously, if you have responded one way to a question and then the correct answer suddenly dawns on you, do change your response. But if you can choose only from a variety of guesses, go with your first or "gut" response.

Take Your Time

Yes, yes, yes, this is easier said than done. But the point is this:

- Plan your time.
- Work efficiently, but not in a panic.
- Make certain that your responses are legible.
- Take time to spell correctly. Even if an instructor does not consciously deduct points for misspelling, such basic errors will negatively influence the evaluation of the exam. Correct spelling of biological terms is especially important—and can be expecially difficult. Be careful.
- Take time to check any mathematical work or chemical formulas. Are the mathematical and chemical formulas correct? Did you calculate correctly?
- Take time to check your graphical work. Biology exams often involve drawing and/or labeling diagrams, charts, or graphs. Make sure your drawings are legible, and make doubly sure that you have labeled all features correctly and clearly. The instructors who have contributed to this book report that they repeatedly observe careless errors of labeling. Points are lost for this.
- If a short answer is called for, make it short. Don't ramble.
- Use *all* of the time allowed. The instructor will not be impressed by a demon-

stration that you have finished early. If you have extra time, reread the exam. Look for careless errors. Do not, however, heap new guesses on top of old ones; where you have guessed, stick with your first guess.

Essay Exams: Read On

Exams in most introductory biology courses are objective, consisting mainly of multiple-choice, fill-in-the-blank, and short-answer questions. A significant minority of survey-level courses, however, also include essay questions or paragraph-length short-answer questions on exams.

"*Essays!* That's for English and history. Biology is about organs, molecules, and populations, not words!"

Well, not always. A growing number of instructors want you to *write* about biological processes and issues. Therefore, please be sure to read Chapter 5 for advice on preparing for and writing effective essay responses.

CHAPTER 4

THINKING LIKE A BIOLOGIST: GEARING UP FOR BIOLOGY

HOW MANY TV COMMERCIALS OR MAGAZINE ADS HAVE YOU SEEN THAT OFFER "SCIentific proof" of the effectiveness of the product being hawked and hyped? That's a powerful phrase, "*scientific* proof."

As a society, we expect a great deal from science and scientists. Without quite being sure why, we equate *science* with truth. We look to science and scientists for answers to technological problems, for cures for dread diseases, for devices to make our lives easier and more pleasurable, and for the answers to the mysteries of the universe. For many of us, science serves the social functions that used to be the exclusive province of religion.

Little wonder, then, that many students eagerly and excitedly embark on their introductory college science courses. And little wonder, too, that even more students face science core requirements with anxiety, trepidation, and even dread.

Science seems so mighty and so mysterious. Or it's for nerds: "eggheads," "propeller heads," *geniuses.*

Scientific Method

You probably don't count yourself among this group. Yet you do have an idea that a key to what scientists do is something called the *scientific method.* Anyone who's taken some high school science has heard about it. Unfortunately, like the general idea of science itself, "scientific method" sounds grandiose, complicated, and mysterious. After all, it's the basis of science!

True, it *is* the basis of science—not just biology, but all science. As different as, say, physics is from biology, both the physicist and the biologist use a similar intellectual approach to guide their investigations. Both use the scientific method.

This must be a profound and complex method indeed!

Not really.

The scientific method is nothing more or less than a systematic approach to a common-sense way of looking at the world—with the emphasis on *systematic.* It works like this:

1. A problem is stated.
2. Facts relating to the problem are gathered.
3. A solution to the problem is proposed: a hypothesis.
4. The hypothesis is tested.

Let's look more closely at these four steps. As the elements of the scientific method, they are the most important first steps you can take toward thinking like a biologist.

The first step is to formulate a problem. For biologists, the problem is typically a question about the order or process of the natural world. Like most scientists, biologists are more interested in the mechanisms by which nature operates than in questions of ultimate purpose. (Such questions are the realm of philosophers and theologians, not scientists.)

Anyone—not just a scientist—can raise questions about the mechanisms of nature. The fact is, however, that most of us don't. Scientists may or may not be *born* curious, but it's a sure bet that they all *develop* curiosity. Without it, step one is impossible. You will increase your chances of excelling in biology—and enjoying the course—if you work at cultivating your curiosity about the world around you. Come to class and approach your textbooks with an attitude of curiosity, and learning will not only be easier, but much more meaningful.

After formulating a problem to study, the next step in the scientific method is to gather facts (data). Facts are gathered by

- firsthand observation
- measurement
- counts
- review of records of past observations

This material must be evaluated for reliability (Is it reliable? Why? Why not? *How* reliable?) and filtered for relevance to the problem. These processes typically require observational skill, an eye for detail, patience, a willingness to work hard, and objectivity. The latter quality is especially important. Data must be approached without prejudice. To prejudge the data is to compromise their usefulness.

Next, an educated guess is formulated concerning the relation between the problem identified and the facts gathered. This is called a hypothesis. Its very special quality is that it *must* be falsifiable—that is, a statement capable of being proven false. It is not necessary that a hypothesis be proven true. Many, many common scientific hypotheses cannot be proven true beyond a doubt, yet they are still useful—as long as they cannot be proven beyond a doubt false.

The next step in the scientific method has as its object, therefore, not necessarily proving the hypothesis true, but proving it false. This is called testing the hypothesis. The scientist does his or her utmost to find ways to contradict or disprove the hypothesis. This is the phase of the scientific method that many people—not just in college, but throughout history—have the most trouble with. It is human nature to hang on to cherished beliefs, even in the face of evidence to the contrary. In the seventeenth century, when the astronomer Galileo introduced observational evidence supporting Copernicus's idea that the earth is not the center of the universe, he was put on trial by the Church, and came perilously close to losing his life. Similarly, ever since Charles Darwin developed the theory of evolution in the mid-nineteenth century, there has been no shortage of individuals and religious organizations who have opposed it.

Why is the theory of evolution called a *theory* rather than a hypothesis? A hypothesis that is repeatedly tested and never falsified is raised to the status of theory. Yet note that the theory of evolution has not been proven true. It has simply never been proven false. It has stood as a sufficient explanation of countless particular instances. It has never been successfully contradicted.

Testing of a hypothesis can be through experiment. An experiment is a test in which the researcher controls as many of the variables—the context and inputs of the test—as possible. Ideally, the effect of altering only one variable at a time is observed and noted. If feasible, a comparison is made between an *experimental group* and a *control group*, which are no different from one another except for the single variable being tested.

Where an experiment is not possible or practical, *observation* (sometimes called *naturalistic observation*) is used. Darwin, for example, did not try to experiment with evolution in a laboratory, but made meticulous observations of the fauna of South America and Africa.

What Is Biology?

Biology is the scientific study of living systems.

Let's look at some of the ground rules of the discipline in order to help you start thinking like a biologist.

The subject of biology—living systems—is terribly ambitious, and, obviously, the subject of life is complex. But biologists typically make it somewhat simpler by approaching it from chemical and physical perspectives. Essentially, they see organisms as living systems that transform matter and energy. This view is called *mechanism*. It holds that life is just a very complex form of physics and chemistry.

Just what is a *living system*, as most biologists define it? It is a physical and chemical unit possessing the following characteristics:

- The system exhibits *metabolism*. That is, the system transforms energy-rich matter into energy and discards energy-poor waste material.
- The system responds to or interacts with its environment. It responds *selectively* to external stimuli produced by this environment. Its selective response generally promotes survival. (For example, a dog will be attracted to the smell of food, but will avoid the heat of fire.)
- The system seeks to maintain *homeostasis* (balance) through mechanisms and processes that reduce or remove harmful conditions. (For example, you sweat in the heat in order to maintain "normal" body temperature.)
- The system grows. Organisms add material to themselves. Even mature organisms continually perform *biosynthesis*, replacing worn-out cells, and so on.
- Living systems reproduce.
- Living systems not only reproduce, but reproduce others like themselves. They carry *hereditary information.*
- Living systems belong to recognizable populations of systems similar to themselves.

NOTE: ***Mechanism*** **is by no means the only possible view of life.** ***Vitalism,*** **for example, holds that life includes something beyond mere physics and chemistry—a kind of vital force.** ***Holism*** **or** ***compositionalism*** **represents a less profound departure from mechanism, but still offers a challenge. While holists agree that life has a chemical and physical basis, they argue that as these elementary components come together in increasingly complex systems, it becomes impossible to study these systems meaningfully if one reduces them to their chemical and physical elements. There is, for instance, no chemical explanation for why a dog should have four legs.**

Beginning biology courses are based primarily on the mechanistic approach, but usually acknowledge the challenge posed by holism.

Even given this very fundamental view of life, the variety of living systems is too great for any single investigator to study. Biologists usually specialize, and such specialization is reflected in the organization of the typical introductory course or two-course introductory sequence, as we shall see beginning with Chapter 6.

Watch Your Language

Just as you must get accustomed to the scientific method in your approach to biology, keeping in mind its four basic steps, so you must understand that biology has a vocabulary all its own. Essential to success in a biology course is learning the vocabulary and then using it *accurately*. This requires memorization, but it also requires care and thought. Typically, scientific terms have limited and precise definitions. In contrast to everyday language, nuance and shades of meaning are avoided. Objective description and labeling are sought.

It is a good idea to create a running list of the technical and specialized words you encounter, complete with definitions *and* examples of *usage.* To be sure, your textbook will have a glossary of terms, and you will also find a basic glossary in this book. But such resources are no substitute for your actually taking the time and effort to create your own list. Don't strive for completeness, but do make note of the terms you most frequently encounter, the terms that are obviously key.

NOTE: In some biology programs, introductory-level courses are not sweepingly broad surveys, but instead concentrate on special aspects of biology, such as the environment or human biology.

While you should use the language of biology with thought and care, you don't have to become paralyzed by fear of misusing terms. Just make certain that you understand each new term you encounter. Don't let any slip by. Such slips are cumulative. If you fail to understand term A, term B becomes that much more difficult to grasp. By the time you arrive at term C, you may be totally in the dark. Remember, too, that the first step toward study of any biological problem or situation is *description,* and biology provides an ample and precisely defined vocabulary to facilitate such description. The specialized terms of this science are the *tools* of the discipline. To work with biology, even at the introductory level, you must have a sound knowledge of the basic tools. It's pretty hard to drive a nail with a screwdriver. Pick up the hammer instead. It's pretty hard to describe *respiration* in a plant if you think the word is synonymous with breathing.

Another List

In addition to your running list of specialized terms, you should compile a list of the major theories, assumptions, and laws that you frequently encounter or that are pointed out to you as especially important. Again, you will certainly find these in your textbook, but writing them down for yourself is a great, *active* way of learning them.

TIP: Think of this list as an annotated index. You should not only write down the particular law or principle, but make a note of where in your textbook it is discussed and explained.

Thinking Graphically

Many laws and principles in biology are not only stated in words, but also in formulas, equations, or tables. Include these in your list. In some cases, graphic representation of the principle is required. For example, the Krebs cycle (which illustrates oxidative metabolism, the main source of cellular energy) is best understood as a diagram. Don't overlook such illustrations. Work at interpreting and understanding them. You may even want to reproduce them in your list of principles.

In addition to diagrams of various sorts, you will also encounter a good many graphs in biology. Resist the temptation to gloss over the graphs you encounter in your text. Instead *study* them—and by *study*, we mean *interpret*: think about the meaning of what you are seeing. The best way to go about this is to look at a graph and try to explain *in words* the relationships illustrated. This is not something you learn to do all at once. It takes practice—just as reading words on a page or notes on a musical score takes practice. There is nothing about C-A-T that *looks* like a cat, but, with practice, we learn to see a cat when we read *cat*. The same principle applies when working with graphs.

Equations—in *Biology*?

Some students choose biology to satisfy a core science requirement because they see it as a nonmathematical alternative to chemistry or physics. While it is true that those disciplines are sure to include lots of work with equations, biology, which encompasses biochemistry as well as calculations relating to such matters as cell mechanics and to organisms populations, does involve some work with equations. And some students find this fact—well, unpleasant.

In lectures and in your textbooks, you will encounter a vocabulary of biochemical, biomechanical, and even statistical equations that are just as essential as the verbal vocabulary we have already discussed. Just as you must make an effort to understand the specialized terms of biology, you should work on understanding each basic equation. These are the building blocks and basic tools of the course.

Fortunately, most chemical and mathematical equations in introductory-level courses are quite straightforward. Just make certain that you take the time to interpret them and understand them. Most instructors suggest that you get into the habit of mentally translating equations into words. Don't just regard them as abstract sets of symbols. They describe important chemical and physical processes.

Pictures

In addition to graphs, diagrams, and equations, biology includes a good many sketches and photographs of organisms and various organic structures. These are not provided just to dress up the text. You should learn all that you can from the sketches and diagrams provided. Exams often call for you to sketch basic organisms—a cell or a cross-section of an earthworm—so be certain that you are sufficiently familiar with the important images in the course.

Lab Work

Many biology courses include lecture sections as well as lab sections or, perhaps, fieldwork. Remember, biology is an *experimental science*. It is highly reliant on direct observation. You can be certain that the work you do in the lab or in the field will show up on exams. Take notes, and be certain that you understand not just the process of what you are doing in the lab, but the significance of it.

Be There

Biology is a complex subject. It is also an *eclectic* subject; that is, it borrows from various allied fields and disciplines, including chemistry, physics, and statistics, and it even overlaps with psychology and medicine. Yet biology is also an *orderly discipline.* It rests on clearly stated assumptions. It has clearly stated objectives and goals.

The key to studying successfully a subject that is complex, eclectic, and orderly is to proceed step by step, and to make certain that you understand each step before you move on to the next. Learn the vocabulary *well.* Understand the basic principles, theories, and laws *well.* Learn to work with graphs, diagrams, equations, and drawings well. Be certain that you understand all lab work fully, process as well as significance.

- Don't skip over textbook material.
- Don't just start reading words without taking the time and making the effort to understand them.
- Don't just glance at the graphs, diagrams, equations, and other illustrations you encounter. Interpret them by translating what you see into words.

Be there. When you study your textbook, *study* it. Practice the concepts you read about. Biology is as much a set of observational skills as it is an intellectual pursuit. Acquiring and developing any skill requires active practice, not just passive reading.

Be there: in class. Don't skip classes. Attend lectures. Take notes. Ask questions. Most instructors believe that *most learning takes place in the classroom.* This means actively listening to lectures, actively participating in class discussions, and—very important—asking questions *as they occur to you.* If you don't understand something in a lecture, ask about it as soon as possible. This is by far the most effective way of learning.

More Than a Set of Problems

We've all known people who are so intensely detail-oriented that, in pursuit of every last fragment of minutiae, they consistently miss the big picture. As the old saying goes, they can't see the forest for the trees. Biology, which requires learning many concepts and a great many new words, unfortunately presents this very trap. You can avoid falling into it by trying to think of biology as something more than the sum of a series of concepts and words and diagrams. These are important, but they alone are not the purpose of biology. You are getting a peek at a field that sets as its goal nothing less than the study of life itself. Try to keep that in mind. It will drive and invigorate your work, and it will help you to bring together all those many constituent parts—vocabulary, tables, graphs, diagrams, drawings, equations—into a meaningful picture of the awe-inspiring mechanics of life.

CHAPTER 5

THE ESSAY EXAM: WRITING MORE EFFECTIVE RESPONSES

SOME EXAMS IN INTRODUCTORY-LEVEL BIOLOGY CONSIST EXCLUSIVELY OF MULTIple-choice and true-false questions. Many more include such questions in addition to fill-in-the-blank segments and one-word or single-sentence short answers. But a growing number of instructors, especially in smaller schools and in introductory courses that concentrate on particular aspects of biology (such as human biology or the environment), include essay questions in their exams. Often, these take the form of short essays, perhaps a paragraph or two in length. In some cases, however, they are more extended. A significant minority of biology instructors writes exams that consist exclusively of essay questions, which may consume several pages of an exam "blue book."

In this book, you'll find examples of all the biology exam types you can expect to encounter.

Downside and Up

Griping is of little value at the outset of any enterprise, including a biology course that not only requires facility with formulas, pictures, and maybe a microscope and dissecting scalpel, but also with words. Nevertheless, griping is human and natural, so we'd better get it out of the way.

Essay exams have the following distinct disadvantages:

1. They are intimidating. Even experienced professional writers may get a shudder when they sit down to a blank page. Where do you begin? Worse, where do

you go once you've begun?

2. Essay questions generally require deeper and broader knowledge of a subject than multiple-choice questions do.
3. Essay exams are time-consuming to take. It may be difficult to budget your time effectively.
4. Essay exams test not only your knowledge of course material, but your language and writing skills. This may seem like an unfair demand.
5. Even in a so-called hard science, essay questions may contain a significant element of subjectivity. Often, the issues are gray rather than black and white. Not only does subjectivity enter into your response, it also plays a role in the instructor's evaluation of the response. An instructor may respond to the skill of the presentation (or lack of such skill) as much as he or she does to the substance of the answer.

In fairness, essay exams are almost as demanding on the instructor as they are on the students. They are much more difficult and time-consuming to grade than "objective" tests are. The instructor who uses essay exams is demonstrating a genuine commitment to her students and her subject.

So there's the downside. But each of these negatives has a corresponding positive—if you know how to find and exploit it. Look:

1. True, your blue book may be blank, but your mind doesn't have to be. First, there are effective ways to prepare for the questions on an essay exam—and we'll talk about these shortly. Second, take a good, long, careful look at the question. It should give you plenty to get you started. Get into the habit of using the terms, parts, and structure of the question as a kind of framework on which you construct the terms, parts, and structure of your answer.
2. If essay questions usually require deeper and broader knowledge of a subject than multiple-choice questions do, they also offer a deeper and broader stage on which you can play out your understanding of the course material. When you respond to short-answer and multiple-choice questions, you are limited by the instructor's rules: true, false, a, b, c, or d. It's a kind of binary situation. Either you *know* the *correct* answer or you don't—and if you don't, you lose. In responding to an essay question, you certainly need to address the question in all of its parts (don't stray, don't evade, don't get off the track), but you have much more control. You can focus on areas you know most about. You can play to your strengths and minimize your weaknesses.
3. Essay exams are time-consuming to take. That's a fact. And your instructor

knows it. He or she takes into account the pressure of limited time and the fact that you are writing a single draft when he or she evaluates the essay. This generally prompts the instructor to overlook a lot of sins of omission and even outright errors. Time pressure actually reduces your instructor's expectations.

4. If essay exams test not only your knowledge of course material, but also your language and writing skills, it behooves you to polish those skills. Good writing will earn extra points. It's that simple. You may not know more about biology than the person sitting next to you, but if you write more effectively, you will earn a higher grade. An added bonus: The more clearly and effectively you can express yourself, the better your own understanding of the material you are writing about will be. Effective writing not only communicates knowledge to others, it helps you to communicate with yourself.
5. Essay questions contain a significant element of subjectivity, it is true, and this can give you that much more "room" to be right. Create a skillful presentation, and you are likely to be evaluated positively, even if you miss some issues.

Study and Preparation

Even if they deny it, most instructors tend to "teach for the midterm and final." More accurately put, they construct exams that genuinely reflect the course content, including particular themes and topics that are emphasized. Do instructors ask "trick questions"? Rarely. Do they deliberately try to mislead you, hiding exam material in the background of the course, as if it were an Easter egg? Almost never. The fact is that instructors want you to succeed. Good test performance tells them that they have gotten through to you. The more students who do well, the more successful an instructor feels. This being the case, be certain to take careful lecture notes. Make a good set of general notes, but also listen selectively for:

TIP: Some instructors prepare examinations well ahead of time, but most write them up shortly before they are given. Usually, the instructor will review his or her notes in preparation for writing an exam. This makes it even more likely that well-emphasized subjects and issues will appear on the exam.

- Points that stand out
- Points that are repeated
- Points preceded by such statements as, "Now this is important" or "This is a major issue" and the like.

Assume that any point, theme, or topic that is given special emphasis will appear as an exam topic. The more emphasis it is given, the more likely it is to appear as an essay exam question.

Typically, course lectures mesh with textbook assignments, additional assigned outside reading, and perhaps class handouts. Take notes on all of these sources. Handouts are usually especially important. Assume that handout material will fig-

TIP: Do not confine your preparation to your notes. If past course exams are available for your review, review them. Use them as practice tests. Of course, you should also make use of the exams in this book. These days, many instructors post past exams on a special web site devoted to the course.

ure in some way on any exam.

Be certain that you know just what part any lab exercises will play in the exam. Essay questions may well focus on lab work, asking you, for example, to describe an experiment, to describe its results, and to draw conclusions as to its significance. Indeed, lab work is a natural for a narrative essay question. Be certain that you take good laboratory notes and that you do not neglect to study them.

Avoid passivity. *Ask* the instructor to talk about the scope of the exam. Also seek out students who have taken the course before. Ask them about the exam. Many instructors have favorite themes or concerns, which get repeated from year to year.

You Are Not Alone

You needn't face the exam alone. Group study is often a highly effective way of preparing for essay exams. Indeed, many instructors encourage students to form study groups. This is because group discussion tends to bring out major themes and issues—the meat and potatoes of essay exams—rather than mere facts, which are the focus of short-answer or multiple-choice exams.

Don't make the mistake of allowing the study group to dissolve into a social hour. Consider focusing the discussion by having each member of the group make up a sample essay question. Use these as the topic of discussion.

Limited Possibilities

Let's assume you're not musically inclined. Now, look at the score of, say, the Beethoven *Moonlight Sonata.* All those notes! All those chords! How does anyone ever learn how to read so much simultaneous information, much less translate it into sound on a keyboard?

PITFALL: By all means, examine tests from previous semesters, but don't make the mistake of assuming that these will simply be repeated this semester. Most instructors change tests from year to year. Examine past tests to get an idea of the type and scope of questions asked—not to get specific answers.

Well, it requires study, hard work, practice—and inborn musical talent helps, too. But it really isn't as hard as it appears to the unmusical. While, theoretically, there is an infinite number of ways in which musical notes can be combined and deployed, in actual practice, the possibilities are indeed limited. Most chords and note sequences occur in recognizable groups and patterns. Just as, when you read a book, you don't spell out individual words or struggle to recognize individual letters, but instead more or less unconsciously interpret familiar linguistic patterns and phrases, so a musician processes the notes he or she sees.

Now, you may think that the range of questions possible in an essay exam is virtually infinite, but, actually, like the notes of a musical score or the words on a novel's page, the range is limited—and this is true regardless of subject.

There are a limited number of questions that can be asked about any theme, subject, or topic. Furthermore, each question is controlled by a key word. Knowing those key words—and understanding their meaning—will help you to prepare adequately for the exam. Here they are:

Analyze Literally, take apart. Break down a subject into its component parts and discuss how they relate to one another.

Compare Identify similarities and differences between (or among) two (or more) things. End by drawing some conclusion from these similarities and differences.

Contrast Set two (or more) things in opposition in order to bring out the differences between (or among) them. Again, draw some conclusion from these differences.

Criticize Make a judgment on the merits of a position, theory, opinion, or interpretation concerning some subject. Support your judgment with a discussion of relevant evidence.

Defend Give one side of an argument and offer reasons for your opinion.

Define State as precisely as possible the meaning of some word, phrase, or concept. Develop the definition in detail.

Describe Give a detailed account of something. Where biology subjects are concerned, this account will typically be step-by-step, with the emphasis on cause and effect.

Evaluate Appraise something, rendering a judgment as to its truth, usefulness, worth, or validity. Support your evaluation with relevant factual evidence.

Explain Clarify something and provide reasons for it.

Identify Define or characterize names, terms, things, places, events, or other phenomena.

Illustrate Provide an effective example of some stated point, principle, or concept.

Outline Show the main features of some event, concept, idea, theory, and so on. Omit the details. Often, such an answer will be in outline rather than narrative form.

Pros and cons A more specialized form of *evaluate*. List and discuss the positives and negatives about a certain position, idea, event, theory, and so on.

Relate Narrate an observation or set of observations; emphasize the relation of one observation to another, especially focusing on cause and effect.

Review Survey a subject. This is much like an *outline*, but put in narrative rather than list or outline form.

State Present your answer briefly and clearly, usually using a simple declarative sentence: *Such-and-such is such-and-such.*

Summarize Give a concise account of major points, ideas, or events. Skip details and examples.

Trace Follow an event, theory, or idea to its origin. The form of this response is generally: "The origin of A is X, Y, Z." Then the rest of the answer continues by elaborating on X, Y, and Z and moves forward to A.

One or more of these are the intellectual operations basic to just about any essay. Look for these key words in the essay question.

Many of these operations focus on just two elements:

- ◆ cause
- ◆ effect

Indeed, most biology essay questions deal with causes and effects. Be prepared in advance to work with both elements.

Test Time: The Problem of Inertia

If you've just finished up an exam in Physics 101, maybe you recall what Sir Isaac Newton had to say about inertia—the tendency of a body in motion to remain in motion and a body at rest to remain at rest. Sitting down to an essay question on exam day, many students are confronted by their own personal form of inertia. You look at the question. And look at it. And look at it some more, hoping, perhaps, that the letters will magically rearrange themselves on the page to yield up the answer.

TIP: You can use the operations and elements just discussed to focus your study notes. For example, instead of listing a bunch of unrelated facts about the subject of photosynthesis, why not focus your notes in terms of causes and effects and also trace particular effects to their causes? You might study photosynthesis by *outlining* all of the elements operative in it, including energy from the sun, the cellular structures involved, the chemical substances and processes involved, and the role of photosynthesis not only in the life of the particular plant in question, but in an entire ecosystem.

The bad news, of course, is that they will do no such thing. But the good news is that the germ of the answer is, in fact, in the question. How do you overcome essay test inertia—that paralyzing difficulty in getting started? Just read the question. *Really* read the question.

In fact, don't worry about writing just now. Sit down and read *all* of the questions before you begin to write anything.

Here's why:

- ◆ If you are given a choice of which questions to answer (say two out of three), you want to be sure that you answer the ones you know best.

- Questions are sometimes related. You want to get an idea of just how they are related to one another, so that you don't "waste" too much of your answer on one question to the neglect of another.
- You need to assess your time needs. Are there some points that will require more time than others?
- You want to make certain that you answer the questions you are confident about first. Given a limited amount of time, be certain that you get to your best shots first and complete them before attending to the questions you're less confident about.

TIP: People who give a lot of speeches are fond of offering this formula for a successful speech—"Tell them what you are going to say. Say it. Tell them what you said." You might keep this in mind when you are answering an essay question, though you should elaborate on the formula a bit:

1. **State your subject or thesis.**
2. **Briefly state how you will discuss it (A, B, C, D—or maybe C, A, B, D).**
3. **Answer the question.**
4. **Concisely summarize your answer.**
5. **Draw any additional conclusions as to significance, ramifications, etc.**

Before you begin to write, be absolutely certain that you understand each question completely. Read each question actively, aggressively, with pencil in hand:

- Identify and underline key words—including those listed above as basic "operations" and "elements."
- Be certain to *do* what the key words ask. For example, don't just *define* when you are asked to *explain.*
- If a question is complex, consisting of several subquestions, make certain you understand and answer all parts of the question.

Generally, you should let the question provide the basic structure for your response. If a question consists of subquestions A, B, C, and D, begin by answering A, then B, C, and D. If you have a *very good* reason for changing the order of your response, be certain to explain it. For example, you might begin: "Because C is essential to understanding A, B, and D, I will begin by discussing C, then proceed to A, B, and D."

Finally, answer the question—and only the question. Make certain that you address all parts of the question, but don't go beyond what the question asks—unless, after you have thoroughly addressed each aspect of the question, you feel it is important to bring in additional issues. If you do so, tell your reader what you are doing, so that he or she won't think you've misunderstood the question and are simply going off on a tangent.

To Outline or Not to Outline

Sometimes, pulling out ideas in response to an essay question is difficult, halting, and laborious. Sometimes, however, you are flooded with ideas. Either situation can make inertia more powerful and, at worst, bring on mental paralysis or, at the very least, cause you to write a poorly organized essay. To prevent these outcomes, apportion your time so that you spend about half of the exam period planning and outlining your response.

Now, an essay exam outline does not have to be a formal outline with major and minor headings. Perhaps a simple list will be sufficient. Just make a map of your answer, setting down the main points that need to be covered. This will have three effects:

1. It will help ensure that you leave out nothing important.
2. It will help you organize the logic of your response.
3. It will reduce your anxiety.

The first two points are obvious. The last is less obvious, but no less important. Without an outline, you may fear that you will forget something important or get lost in your response. Get the main points out of your head and onto paper quickly, and you won't have to worry about forgetting anything or getting lost.

Structure Strategy

Make the structure of your essay exam response as clear and obvious as possible. Begin with a thesis statement: a statement or listing of the main idea you will support and develop in the body of the essay.

Where do you get your thesis statement?

The first place to look is the question: "Species diversity is a major area of study in ecology. What is species diversity? Is there a possible relationship between species diversity and the stability of communities?"

There's the question. Begin by getting to the point: "Species diversity is a subcategory of biodiversity that refers to the number of species that make up a community."

Now, it is the second question—"Is there a possible relationship between species diversity and the stability of communities?"—that provides you with the thesis statement that will drive your essay. Just convert the question into a statement: "Species diversity often aids in stability."

This thesis statement should naturally lead you to explain how you intend to support it: "A prime example of how species diversity aids stability is when a community is attacked by a force such as disease. As we saw in [*example from lecture*], the mix of species in a community may prevent the disease from wiping out all members of one or more species."

From this point, the essay should go on to elaborate on the example, giving the details of the particular example. And that is the body of this simple essay. It is essentially a presentation of material acquired in lecture, but driven and *focused* by a thesis statement *directly derived* from the question.

In general, make the thesis statement as simple and as direct as possible. State the thesis. Present your plan for supporting your thesis. In this case, your "plan" is also a simple one: you propose to present a "prime example" of how species diversity enhances community stability.

Follow your plan in the body of the essay:

I. Thesis: "Species diversity often aids in stability."

II. Plan: "A prime example of how species diversity aids stability is when a community is attacked by a force such as disease."

III. Example 1: "As we saw in [*example from lecture*], the mix of species in a community may prevent the disease from wiping out all members of one or more species."
   A. Narrate the example.
   B. Highlight a way in which the example shows how species diversity contributes to community stability.
   C. Highlight another way.
   D. And another.

IV. Example 2 (perhaps from the textbook)
   A. Narrate the example.
   B. Highlight a way in which the example shows how species diversity contributes to community stability.
   C. Highlight another way.
   D. And another.

V. Example 3 (Maybe you have a third example from your outside reading.)
   A. Narrate the example.
   B. Highlight a way in which the example shows how species diversity contributes to community stability.
   C. Highlight another way.
   D. And another.

VI. Conclusion
   A. Compare and contrast the two or three examples.
   B. How do the two or three—in differing ways—confirm the thesis (that species diversity enhances community stability)?

Of course, if you have only the one example, omit the comparison and contrast and just make a concise statement about how the example demonstrates that species diversity enhances community stability.

The K.I.S.S. Formula

Can you really get ahead by giving your professor a KISS? Well, sort of. This acronym stands for *K*eep *I*t *S*imple, *S*tupid.

Now, let's get something straight. This does not mean that you should overly simplify complex ideas or issues, let alone avoid them. But it does mean that you should structure your presentation of ideas in as simple a form *as possible.* This means:

- Be concise.
- Be direct.
- Start the essay with a thesis statement.
- Start each paragraph with a topic sentence, announcing the subject of the paragraph.
- Try to make a single major point in each paragraph.
- When you move on to a new point or new idea, start a new paragraph. Essay exam essays usually have short paragraphs—certainly shorter than what you'd write in a term paper, for example.
- Draw a definite conclusion.

Specify, Always Specify

Wherever possible, avoid abstraction. In place of vague generalizations, make very specific points that use specific examples. Examples are important in any essay response, and they are especially important in essays on biological subjects. Remember, science begins with observation—that is, science *begins* with examples.

Signposts

Develop a repertoire of verbal signposts. One of the most effective signposts is enumeration. For example, instead of saying "The scientific method consists of *several* steps," write "The scientific method consists of *four* major steps." Then go on to list and discuss all four steps.

Enumeration accomplishes three things:

1. It creates the impression that you are in control of the information.
2. It creates an impression of precision and completeness.
3. It sets up an expectation in the reader, who is satisfied when that expectation is fulfilled. You promise four items. You deliver four items. The reader is impressed.

Other signposts include words and phrases such as:

To begin with . . .

First . . .
Next . . .
Therefore, . . .
If . . . then
Because . . .
The result of . . .
. . . caused by . . .
However . . .
Except for . . .
Including . . .
For example, . . .
Although . . .
Since . . .
Consequently . . .
Finally, . . .
In conclusion . . .

TIP: The language of biology is typically very precise. Make certain that you use it correctly. For example, in everyday speech, we might use the word *respiration* as a synonym for breathing. To a biologist, however, respiration is the oxidative process within the cell by which the chemical energy of organic molecules is released through the metabolic process. Quite a difference!

Use these to get your reader from one point to the next, to make clear exceptions, and to point your reader toward your conclusions.

A Few Words on Words

Take time to choose your words carefully. This does not mean trying to impress the instructor with big words or fancy words, but do try to find the right words, the words that most precisely say what you mean.

- Use the language of biology. Identify and become comfortable with the specialized terms used in lectures and textbooks. Understand them thoroughly. Use them appropriately in writing the exam essays.
- Avoid slang. Slang is not only imprecise, it creates a poor impression.
- Prefer strong, precise nouns and verbs to adjectives and adverbs. This will help you to convey more accurate meaning. From the point of view of a biologist, the sentence "Chitin is the hard substance that exoskeletons are made of" is a significantly weaker statement than "Chitin is a protein-polysaccharide complex found in exoskeletons."

TIP: Try to budget time to reread and proofread your essay. Catch and correct errors of usage, grammar, and spelling.

- Express yourself in a direct manner. Don't load up sentences with unnecessary words. Try to make each word count. Use the active voice instead of the passive voice: "Rapidly beating cilia move the cell from place to place" is a much stronger sentence than "The cell is moved from place to place by rapidly beating cilia."

- Avoid unnecessary qualifying phrases and waffling words such as "it has been said" or "I think" or "it seems to me" or "it seems likely that." Make direct statements.
- Avoid padding and repetition like this: "Atoms are very small. In fact, they are the smallest part of an element—if you don't count the subatomic particles that make up the atom." Instead: "Atoms are the smallest whole particle of an element."
- Avoid errors of usage, grammar, and spelling. If you have trouble in these areas, work on them. Such errors undercut your credibility.

Neatness Counts

Take enough time to write legibly. Make your work as easy to read as possible. If the instructor has to struggle to decipher your handwriting, he or she will easily lose the thread of your discussion, and your grade is likely to suffer as a result.

Recycle!

When the graded exam comes back to you, resist the temptation either to pat yourself on the back or to kick yourself in the backside. Instead, carefully read the examiner's comments. Learn from them. Schedule a conference to discuss the exam—not with the goal of getting your grade changed, but of identifying those areas that can use improvement.

When you have a conference, try not to respond defensively. Invite frank feedback. Don't get offended or upset. Instead, look for patterns. Does the instructor say that you simply failed to answer the question adequately? That you didn't answer all parts of the question? That you answered vaguely? That you punctuated poorly? That you didn't use enough examples? Diagnose areas that need improvement, even as you identify your strengths. Your ultimate goal is to avoid repeating errors while working to duplicate your successes.

TIP: Some biology essay questions call for you to illustrate your answer. You don't have to produce great works of art, but be sure that your illustrations are legible, with all features clearly—and correctly—labeled. Slow down. Take enough time to be neat.

PART TWO

STUDY GUIDE

CHAPTER 6

INTRODUCTION TO GENERAL BIOLOGY: THE MAJOR TOPICS

BIOLOGY, WE SAID IN CHAPTER 4, IS THE SCIENTIFIC STUDY OF LIVING SYSTEMS. MORE specifically, it studies living things as well as life itself, with emphasis on the mechanisms by which life exists and functions. Obviously, this is a vast field. Biologists study life in its smallest, simplest forms—the viruses and unicellular organisms—up through forms of increasing size and complexity, including human beings, and, beyond this, to entire populations, limited only by the extent of planet Earth. Indeed, some biologists work with astronomers and others to speculate about what life might be like beyond our own planet. Biologists also study life at its most basic chemical and physical levels as well as at more complex anatomical, physiological, and behavioral levels, up through the ecology of entire populations.

Landmarks

Biology is not only an expansive science, it is also a very old one. Its origins are usually traced to the ancient Greek philosophers, primarily Hippocrates (460-370 B.C.), the fifth-century B.C. "Father of Medicine," and Aristotle (384-322 B.C.), who lived in the fourth century B.C. Another Greek, Theophrastus (380-287 B.C.), is remembered as the Father of Botany, because it was he who first attempted to organize the study of plants. Galen (A.D. 130-200) was a Greek physician who thoroughly (though not always accurately) studied animal and human anatomy.

As is true of many sciences, biology made few advances during the long "Dark

Ages," and it was only with the Renaissance and the work of Andreas Vesalius (1514-1564) that Galen's time-worn anatomy was at last challenged and refuted. The work of Vesalius is especially important because it was based on actual observation, through dissection of cadavers, rather than on lore accepted at secondhand. While most introductory courses are not extensively concerned with the history of the discipline, you will almost certainly encounter these additional landmarks in the development of biology:

- **Robert Hooke** (1635-1703) first describes (and names) "cells" (in cork).
- **William Harvey** (1578-1667) demonstrates circulation in the human body.
- **Francesco Redi** (1626-1697) disproves "spontaneous generation"; shows that life comes from life and only from life. His work is often cited as the "first" biological experiment.
- **Anton Von Leeuwenhoek** (1632-1723) uses the newly invented microscope to study living cells.
- **Carolus Linnaeus** (1707-1778) devises "binomial nomenclature," the basic system of classification of animal and plants still used today.
- **Luigi Galvani** (1737-1798) experiments with "animal electricity"—the beginning of nerve physiology.
- **Joseph Priestley** (1733-1804) is first to study photosynthesis.
- **Jean Baptiste Lamarck** (1744-1829) coins the word biology.
- **Charles Darwin** (1809-1882) develops the theory of evolution by natural selection.
- **Louis Pasteur** (1822-1895) explains fermentation and pioneers bacteriology.
- **Gregor Mendel** (1822-1884) becomes the father of genetics.
- **1910s:** genetic mapping technique is developed.
- **1950-1960s:** the structure of DNA, in which heredity is molecularly encoded, is described.

The Branches of Biology

Modern biology is far too vast a field for any one person to master. It is typically divided into two broad areas, *botany*, the study of plants and plant life cycles, and *zoology*, the study of animals and their life cycles and life histories. Indeed, some universities no longer have a "Biology Department," but instead support separate Botany and Zoology departments (and sometimes other specialized life sciences departments as well).

Within both botany and zoology are numerous subspecialties. Botany is traditionally divided into such subspecialties as

- **Bryology** (the study of mosses and liverworts)
- **Mycology** (the study of fungi)
- **Paleobotany** (the study of fossil plants)
- **Pteridology** (the study of ferns)
- **Palynology** (the study of pollens)
- **Plant pathology** (the study of plant disease)

Traditional zoological subspecialties include such fields as

- **Entomology** (the study of insects)
- **Ethology** (the study of animal behavior)
- **Herpetology** (the study of reptiles and amphibians)
- **Invertebrate zoology** (the study of animals without backbones)
- **Ichthyology** (the study of fishes)
- **Mammalogy** (the study of mammals)
- **Ornithology** (the study of birds)
- **Vertebrate zoology** (the study of animals with backbones)

The specialties and subspecialties above adequately describe traditional biology, but the field has expanded so rapidly in the twentieth century that relatively new specialties have appeared. The most important of these include:

- **Molecular biology** or **biochemistry**, the study of life processes at the molecular level
- **Morphology**, the study of plant and animal structure; morphology is usually further divided into anatomy (study of gross structure), histology (study of microscopic structure on the tissue level), and cytology (study of microscopic structure on the cellular level)
- **Taxonomy**, the science of classification; within this specialty is *systematics*, the identification and ranking of all plants
- **Embryology**, the study of embryo development
- **Genetics**, the study of heredity and the processes of heredity
- **Evolution**, the study of origins of living things and the relationships among them through time
- **Paleontology**, the study of ancient life through fossil records
- **Ecology**, the study of how plants and animals relate to one another and to their physical environment

Basic Assumptions, Wide Variation

As diverse as the many fields of biology are, they are unified by the principles of the scientific method discussed in Chapter 4 and by the basic assumptions about living systems also enumerated in Chapter 4 (metabolism, selective response to the environment, homeostasis, growth and biosynthesis, reproduction, heredity, membership in populations). Yet you will find that biology departments within various universities and colleges take different approaches to introductory biology. Many do offer a general survey of the entire field, often spreading the material over a two-semester sequence. Other departments take a more specialized approach to introductory courses, introducing students to the principles of biology through the study of human biology, ecology, or even molecular biology. The sample exams in Chapters 19 through 29 come from introductory courses of both varieties, some general surveys and some more specialized, reflecting the way the field is introduced in different programs.

The Biochemical Front Door

As many biology departments, biology instructors, and biology textbooks see it, the most logical entry into the study of biology is at the most elemental level, the level not of entire plants or animals, but of atoms and molecules, the level of biochemistry. Depending on the instructor, the course syllabus, and the level of your science background, this may be intimidating. Chemistry is, of course, a complex subject, and organic chemistry is among the most challenging branches of chemistry. Fortunately, most introductory biology courses and textbooks attempt to keep the chemistry they present pretty basic. Many introduce it as part of a discussion of the physiology of the cell, which serves to make the chemistry portion of the course more immediately relevant to the concerns and interests of the *introductory* biology student. Still, it can be a shock to expect to dive into the world of plants and animals only to find yourself juggling atoms and molecules. Be prepared.

Memory

Also be prepared to memorize. Some introductory biology courses go out of their way to make the subject more “relevant” to the nonscience major, linking biology to such topics as human health and nutrition as well as to ecology and environmental responsibility. Even these courses, however, involve learning—and committing to memory—a good many concepts, terms, and even basic equations. The more traditionally organized general biology courses require even more memory work, including mastery of some aspects of taxonomy, the science of classification of plants and animals.

While most biology instructors emphasize learning concepts over simply memorizing facts, there *are* plenty of facts to memorize, as will be apparent if you even skim some of the multiple-choice and short-answer exams included in this book. Whatever else biology is, it is based on description and classification of the natural world around us, and a portion of that data is part of all introductory-level courses.

Road Map

Remember, this book is a guide to help you use midterms and finals, inevitable facts of college life, to focus study of your biology course. It is not a comprehensive introduction to biology, and it is certainly not a substitute for reading your textbooks, attending lectures, and actively participating in class discussion. This chapter and the others in this part of the book should serve to point out the most prominent biology landmarks you will encounter on your trip through the course. We have already inventoried the most basic assumptions and principles of biology. Here are the major themes and topics that are built on these assumptions and principles and that are reflected in the next twelve chapters.

The Chemistry of Life

- Elements, Atoms, Molecules
- Key Functional Groups
- Key Compounds
- Carbohydrates
- Lipids
- Proteins
- Nucleic Acids
- Enzymes, Cofactors, and Vitamins

The Cell

- Cell Variety and Unity
- Cell Characteristics
- Anatomy of a Cell
- Membrane
- Cytoplasm
- Endoplasmic Reticulum
- Ribosomes
- Golgi Body
- Lysosomes

- Peroxisomes
- Glyoxysomes
- Mitochondria
- Nucleus
- Plant and Animal Cells: Some Important Differences
- Tissues: The Next Level of Organization
- Respiration
- Anaerobic Phase
- Aerobic Phase
- Oxidative Phosphorylation
- Cellular Division
- Mitosis and Cytokinesis
- Meiosis

Taxonomy

- The Species Concept
- Phylogenetic Taxonomy
- How Many Kingdoms?
- Purpose of Taxonomy

Bacteria, Viruses, Viroids, and Prions

- The Prokaryote World
- The Archaeobacteria
- The Eubacteria
- Cyanobacteria
- Viruses
- Viroids and Prions

Protists and Fungi

- The Protists
- Protozoa
- Funguslike Protists
- Plantlike Protists
- The Fungi
- Fungi Types
- The Lowly Fungi?

The Photosynthetic Plants
- Photosynthesis
- The Lower Plants
- Chlorophyta
- Phaeophyta
- Rhodophyta
- The Higher Plants
- The Root
- The Stem
- The Leaf
- Bryophytes
- Tracheophytes
- Psilopsida
- Lycopsida
- Sphenopsida
- Pteropsida
- Spermopsida

Simple and Complex Invertebrates
- Distinctive Animal Characteristics
- Phylum Porifera
- The Coelenterates
- Platythelminthes
- Nematoda
- Annelida
- Mollusca
- Arthropoda
- The Importance of Invertebrates

The Chordates
- Echinodermata
- Hemichordata
- Tunicata and Cephalochordata
- The Vertebrates
- Agnatha
- Chondrichthyes

- Bony Fish
- Amphibia
- Reptilia
- Aves
- Mammalia

Animal Systems: An Overview

- Digestion
- Respiratory System
- Circulatory System
- Excretory System
- Skeletal System
- Muscular System
- Nervous System
- Endocrine System
- Pituitary Gland
- Thyroid Gland
- Parathyroid Glands
- Adrenal Glands
- Pancreas
- Ovaries and Testes
- Pineal Gland
- Thymus Gland
- Other Endocrine Secretions
- Reproductive System
- The Male Reproductive System
- The Female Reproductive System

Genetics

- Classical Genetics
- The Gene Concept
- Punnett Squares
- A Loophole in Mendel's Law of Dominance
- Crossovers and Mutations
- Sex-Linked Traits
- Action of Multiple Alleles
- Molecular Genetics

Evolutionary Theory

- Two Theories of Evolution
- Beyond Darwin
- Mutation
- Gene Flow and Genetic Drift
- Natural Selection Revisited
- A Matter of Time

Ecology

- Thinking in Terms of Ecosystems
- Niches and Competition
- Another Way of Looking at the Ecosystem
- Food Chains
- Abiotic Cycles
- The Limits
- Succession
- Biomes
- The Human Element

CHAPTER 7

THE CHEMISTRY OF LIFE

LIFE TAKES VASTLY VARIED FORMS, BUT AT THE MOST BASIC LEVEL, THAT OF CHEMICAL organization, life has a remarkable sameness about it. All introductory general biology courses include some chemistry, and many begin with the chemical basis of life.

Elements, Atoms, Molecules

If you are taking a course for nonscience majors, you will probably start at the very beginning, with a review of the concepts of *elements* and *atoms.* All things, living or nonliving, are composed of elements, of which more than a hundred exist. Oxygen, nitrogen, calcium, sodium, hydrogen, and carbon are examples of elements that are especially important to organisms.

Each element is composed of a particular kind of atom, the smallest constituent of an element that can enter into combinations with atoms of other elements to form *molecules.* It is important to understand how atoms are structured, since this determines how they can combine to form the molecules that are the basic building blocks of life.

Each atom consists of a *nucleus,* which contains *protons* and *neutrons.* Orbiting the nucleus are electrons. Protons are positively charged, neutrons have no charge, and electrons are negatively charged. The electrons are arranged in *orbitals* or *shells* at different energy levels and distances from the nucleus. Atoms are most stable when their outer shells contain a full complement of electrons. Atoms tend to lose or gain electrons until the outer shells achieve a stable arrangement. Gaining, losing, and sharing electrons is the basis for chemical reactions and the bonding of atoms into more complex molecules.

A molecule is a specific arrangement of atoms from the same or different elements. Ozone, O_3, is an example of a molecule made up of three atoms of the same element, oxygen. Water, H_2O, is an example of a molecule made up of atoms from different elements: two hydrogen atoms and an atom of oxygen.

Molecules may come together to form compounds. In order to form molecules, atoms must bond with one another. Electrically charged atoms (that is, atoms with either a surplus or a deficit of electrons) form *ionic bonds.* Metallic elements tend to lose electrons (which are negatively charged) and form *positive ions.* H^+, hydrogen, is an example of a positive ion. Nonmetallic elements tend to gain electrons, forming *negative ions,* such as Cl^-, chloride. Some atoms, including hydrogen, oxygen, nitrogen, and carbon—all highly important in the chemistry of life—also share electrons to form *covalent bonds.*

Ionic compounds form when positive ions (called *cations*) and negative ions (called *anions*) attract one another electrically and bond. When electrons are shared, covalent bonds are created. Carbon forms covalent bonds readily and is the basis of all *organic compounds.*

Key Functional Groups

Life is said to be carbon-based because this readily bonding element is central to all organic compounds. Carbon atoms can bond covalently into long chains and other shapes in order to create the complex molecules of which life is "built." Two of the most basic organic compounds are the hydrocarbons, which are made up of various arrangements of hydrogen and carbon, and carbon dioxide (CO_2), which is produced by organisms as a product of respiration and is used by plants in the important energy-conversion process of photosynthesis.

In addition, you will encounter certain functional molecular groups, which form key classes of organic compounds. The most important include:

- The alcohols (—OH)
- The aldehydes (—CHO)
- The ketones (—CO—)
- The acids (—COOH)
- The amines (—NH_2)
- The amino acids (which include —COOH and —NH_2)

Key Compounds

Four important groups of compounds are associated with living things: carbohydrates, lipids, proteins, and nucleic acids.

Carbohydrates

Almost all organisms use carbohydrates as sources of energy. Some carbohydrate compounds also serve organisms as structural material. There are two broad categories of carbohydrates, simple and complex.

The simple carbohydrates are commonly called sugars, and may be monosaccharides (consisting of single molecules) or disaccharides (composed of two molecules). Probably the single most important sugar you will encounter is *glucose*, a monosaccharide ($C_6H_{12}O_6$) that is the basic "fuel" of living things, the starting point of respiration and the principal end product of photosynthesis.

Complex carbohydrates, called polysaccharides, are made up of linked monosaccharides, and include (among many others) the *starches*, *glycogen*, and *cellulose*.

Starches serve organisms as a storage form for carbohydrates and are, therefore, a highly important category of food. Rice, wheat, corn, and potatoes are all starch sources.

Like the starches, glycogen is composed of many units of glucose, but these are bonded in a different pattern than they are in the starches. With organic compounds, the patterns in which atoms and molecules are bonded is of great importance in how the compound functions. Glycogen is how the human liver stores glucose.

Yet another carbohydrate, cellulose, serves not principally as a source of energy (food), but as structure. Like the other carbohydrates we have just discussed, cellulose consists of glucose units, but they are bonded in patterns that cannot be readily broken down as food (except by a few species of organisms—termites, for example). The walls of plant cells consist of cellulose, as does wood.

Lipids

Organic molecules consisting of carbon, hydrogen, and oxygen atoms, lipid compounds include *steroids* (the basis of many hormones), *waxes*, and *fats*. Fats serve structural purposes as well as store energy.

Proteins

Proteins are among the most important and most complex of organic compounds. They are composed of *amino acids* (molecules consisting of carbon, hydrogen, oxygen, and nitrogen atoms, with, in some cases, other elements as well). While protein molecules can be very large and complex, they all consist of long chains of amino acids (of which there are twenty kinds). The bonds between amino acids forming proteins are called peptide bonds.

Proteins play a central role in all life. They are the principal molecules forming the structure of organisms. They are found in the *cytoplasm*—the watery substances—of cells, and they also are key constituents of bones, tendons, ligaments, and other tissue.

Proteins also form *enzymes*, which are essential catalysts in basic chemical processes within organisms. As catalysts, enzymes are not consumed in these chemical reactions, but they play an essential role in these reactions by accelerating them.

Finally, proteins also serve the organism as a reserve energy source within each cell.

As is apparent from this brief discussion, proteins are central to virtually every aspect of an organism: structure, chemical processes, energy. Proteins may also be seen as responsible for giving each species of organism its unique shape, structure, and identity. That is, each species manufactures proteins unique to the species. How does the particular organism "know" how to do this? A genetic code unique to each species specifies the sequence of amino acids in that species' proteins.

Nucleic Acids

That genetic code is contained within the structure of the fourth group of organic compounds, the nucleic acids. *Deoxyribonucleic acid (DNA)* is the constituent molecule of genes, the bearers of hereditary traits from parent organism to offspring. *Ribonucleic acid (RNA)* controls how an organism builds protein. RNA carries information from the genes in the cell nucleus to structures in the cell cytoplasm, "instructing" them how to carry out various biochemical processes. These matters will be surveyed further in Chapter 15.

Enzymes, Cofactors, and Vitamins

This brief overview has surveyed the major areas of biochemistry to which you are likely to be introduced at or near the outset of a beginning course in general biology. There is one additional class of proteins that should be considered separately. We've already mentioned enzymes, organic molecules that speed up chemical reactions without themselves being consumed or altered. To do their work, some enzymes require small amounts of nonprotein components, metallic ions called *cofactors*. Some enzymes require *organic* cofactors, called *coenzymes*. Both the metallic ion cofactors and the organic coenzymes are familiar as the food supplements many of us take. Examples of metallic ion cofactors are iron and magnesium; and the fat-soluble vitamins (A, D, E, and K) and water-soluble vitamins (C, B_1, B_2, B_3, and others) are familiar examples of organic cofactors—coenzymes.

CHAPTER 8

THE CELL

THE CHEMICAL PROCESSES BRIEFLY SURVEYED IN THE PRECEDING CHAPTER TAKE place within the basic units of organisms, the cells. Just as the same basic chemistry and chemical processes are common to all living things, despite the great variety of life, so the cell is common to all organisms. (Viruses, which are discussed in the next chapter, share some of the properties of organisms, but are not living things.) Some very simple organisms consist of a single cell (they are *unicellular*), but most organisms are multicellular.

Cell Variety and Unity

Like other living things, cells exhibit a great range of variety. They differ in shape (some are round, rectangular, elongated, tapered, spherical, and so on), in size (ranging from about 5 to 50 micrometers—.005 to .05 millimeters), and in function. Some cells are capable of movement from place to place (they are *motile*), some are not. Some perform many functions, while others specialize. For example, a few of the specialized cells in the human body include sensory nerve cells, motor nerve cells, rod cells (in the retina of the eye), cone cells (also in the retina), sperm cells, egg cells, gland cells, and so on. In general (although there are many exceptions), the membranes of plant cells are enveloped by rigid walls, whereas the membranes of animal cells are more delicate and flexible—some can even change shape.

Cell theory states that cells are the basic units of life, that all organisms—plants and animals alike—are made of cells, and that cells arise only from existing cells.

Despite the vast variety of cells, all share four major characteristics, and all share similar parts, called *organelles.*

Characteristics

The four characteristics of all cells relate to:

- **Structure:** Cells are the basic unit of structure in plants and animals.
- **Function:** Whatever the specialized function of a given cell may be, all cells are living units. They may live independently as a unicellular organism, or they may live as part of some animal or plant tissue. As a living unit, each cell is a biochemical factory, which transforms food molecules into energy and for growth, biosynthesis (repair or replacement of damaged or aged tissue), and reproduction. This transformation process is called *metabolism.* All cells perform this function.
- **Growth:** Each cell is a unit of growth. While some organisms remain unicellular throughout their entire life span, even the most complex multicellular organisms begin life as a single cell. Such organisms grow through the orderly increase in the number of cells.
- **Heredity:** New cells arise from existing cells. A cell grows, then divides, producing two cells identical to itself. Cells carry hereditary information from one generation to the next.

Anatomy of a Cell

In essence, a cell consists of a fluid called *cytoplasm* surrounding a *nucleus* and enclosed within a *membrane.* Within both the nucleus and the cytoplasm are additional parts.

Membrane

The cell membrane not only defines the boundary of the cell, but controls passage of materials into and out of the cell. Semipermeable, it behaves in a highly selective way to regulate what goes into and what leaves the cell (this is called *transport*). You may spend considerable time, even in an introductory course, studying the structure and function of the cell membrane.

Cytoplasm

Cytoplasm lies within the cell membrane, but outside of the cell nucleus. The cytoplasm supports the various organelles as well as a system of exceedingly fine microtubules and microfilaments (the *cytoskeleton*), which both support the cell and direct the flow of materials within it.

Endoplasmic Reticulum

This fine network of membranes consists of tubes and flattened sacs spread throughout the cytoplasm. This structure is involved in the circulation of materials within the cell and also synthesizes some of the materials required by the cell.

Ribosomes

These very small organelles dot parts of the endoplasmic reticulum. There are thousands in each cell. Ribosomes are the sites of protein synthesis.

Golgi Body

Also called the Golgi apparatus, this organelle stores, "packages" (wraps in membranous material), and transports proteins and lipids synthesized by various parts of the cell. Plant cells may contain hundreds of Golgi bodies, while animal cells typically have ten to twenty.

Lysosomes

Lysosomes are special *vacuoles* (cavities) containing a digestive enzyme essential to digesting food for the use of the cell. In specialized cells such as the white blood cells of the human body, lysosomes engulf and destroy invading bacteria and thus are a key element of the body's immune system. If, for some reason, the lysosome ruptures, the cell itself will be destroyed in a process called *autolysis.*

Peroxisomes

Like lysosomes, peroxisomes can serve to defend the cell against invading substances. But whereas the lysosome engulfs and digests material, the peroxisome engulfs toxic substances, rendering them harmless by adding oxygen to them.

Glyoxysomes

These organelles, commonly found in plant seedlings, contain enzymes that convert fatty acids to carbohydrates.

Mitochondria

Organelles second only to the cell's nucleus in size and importance, the mitochondria (singular, mitochondrion) are the sites of cellular respiration; in effect, they are the power houses, the furnaces of the cell, producing the energy the cell needs to carry out its many functions.

Nucleus

While the nucleus is the largest and most prominent feature in most cells, a few bacteria and algae species lack this organelle; such cells are called *prokaryotes.* Cells with nuclei are called *eukaryotes.*

The nucleus is composed mainly of protein and DNA, the DNA organized into *chromosomes* (which will be discussed further, below and in Chapter 15). Within the nucleus are two spherical bodies called *nucleoli,* in which RNA is synthesized and stored. The entire nucleus is encased within a double membrane, which has pores that communicate with the rest of the cell.

Plant and Animal Cells: Some Important Differences

The single greatest difference between the typical plant and the typical animal cell is that a rigid wall of nonliving cellulose surrounds the membrane of a plant cell, whereas the membrane of an animal cell is free and usually flexible. The typical plant cell also contains a large vacuole, a space in the cytoplasm that is filled with *cell sap,* the nutrient substance of the cell. In contrast, the storage areas of animal cells are quite small.

Plant cells also contain *plastids,* structures that float in the cytoplasm and hold pigment molecules or starch. *Chloroplasts* are specialized plastids containing *chlorophyll,* which not only imparts to green plants their color, but which can trap light and convert it into the energy that drives *photosynthesis,* a foodmaking process.

Animal cells often have *flagella* and/or *cilia,* threadlike projections of cytoplasm that propel cells capable of movement and that sometimes serve other functions, such as catching food or removing substances harmful to the organism.

Tissues: The Next Level of Organization

Groups of similar cells, which are related in function and location, are called *tissues.* Tissues represent the next highest level of organization above that of the individual cell. Tissues are then grouped into an even higher level of organization as *organs.* Organs that perform related functions are grouped into *systems.*

There are six basic types of animal tissue:

1. **Epithelial tissue** covers and lines organs, providing protection.
2. **Nervous tissue** is made of cells called *neurons,* which specialize in carrying nerve impulses. *Motor neurons* carry impulses to muscles and glands to stimulate movement and glandular activity. *Sensory neurons* carry information from sense organs to the spinal cord and brain. *Interneurons* transfer impulses from sensory to motor neurons.
3. **Muscle tissue** responds to impulses from motor neurons. *Smooth muscle tissue*

makes up the involuntary muscles, such as the muscles of the small intestine, the operation of which is automatic and not under control of the individual. *Cardiac muscle* is peculiar to the heart. *Striated muscle* makes up the muscles that are controlled by the individual.

4. **Blood** is specialized tissue in a plasma medium. The major constituent cells of this type of tissue are *erythrocytes* (red blood cells), which carry oxygen through the organism, and *leukocytes* (white blood cells), which fight invasion by disease organisms. In addition, *platelets* are cell fragments, which are important in clotting the blood.
5. **Connective tissue** consists of nonliving material surrounding living cells. Cartilage, ligaments, and tendons are examples of this tissue type.
6. **Bone**, like connective tissue, consists of living cells as well as nonliving matrix material.

Plant tissue is of three main types:

1. The **epidermis** is a layer of cells covering the surface of leaves, stems, and roots. The epidermis protects the tissues beneath it.
2. **Vascular tissue** conducts material through the plant. *Xylem* conducts water and dissolved nutrients upward from the plant roots. *Phloem* transports food materials to all parts of the plant.
3. **Fundamental tissue** makes up most of the body of the plant. There are three major types of fundamental tissue. *Parenchyma*, which contains plastids, including chloroplasts, is the site of photosynthesis. *Sclerenchyma*, consisting of living as well as dead cells, provides the mechanical support for plant stems and the covering of such structures as seeds. *Collenchyma* are living stem cells that provide additional support to the plant.

Respiration

Cells, individually as well as collectively as tissue (and organs and systems), do a tremendous amount of work, which requires a constant flow of energy. The cell transforms glucose into energy through a process called *respiration*. There are two phases, *anaerobic respiration* (also called glycolysis) and *aerobic respiration*, plus a final step in aerobic respiration, *oxidative phosphorylation*.

Anaerobic Phase

In the cytoplasm, a molecule of glucose is activated by energy supplied by adenosine triphosphate (ATP), a molecule in which the cell stores a "packet" of energy. Through the operation of enzymes, the glucose is converted to pyruvic acid. During this process, energy is concentrated as hydrogen bonds are broken. Ultimately, this energy is released to molecules of adenosine diphosphate (ADP). Whenever the ADP accepts energy, inorganic phosphate from the cellular fluid becomes attached to the ADP molecule, upgrading it to ATP. Thus the process of anaerobic respiration is partially cyclical.

Aerobic Phase

The site of aerobic respiration is the mitochondria of the cell. The pyruvic acid produced by the anaerobic phase of respiration is converted to a compound called acetyl coenzyme A. Through a complex cycle of chemical changes (called the *Krebs cycle*), "fuel" molecules are broken down, releasing in the process a large amount of energy, which is stored in ATP "packets." Atoms of hydrogen, from compounds formed during the anaerobic and aerobic phases, then enter the final phase of respiration, oxidative phosphorylation.

Oxidative Phosphorylation

Another coenzyme, nicotinamide adenine dinucleotide (NAD), receives the hydrogen atoms, which are passed along an electron transport chain that is part of the membrane of the mitochondrion. As the hydrogen atoms are moved along, the hydrogen electrons are passed down the chain, but the hydrogen protons are pushed to the outside of the mitochondrial membrane. This establishes an electrochemical gradient across the membrane, which supplies energy that forms more ATP molecules, or energy "packets." (The ATP is created when inorganic phosphate joins chemically to ADP; hence the name "oxidative phosphorylation" for this final phase of aerobic respiration.) Finally, at the end of the transport chain, the hydrogen electrons rejoin the hydrogen protons and combine with oxygen to form water.

Step Back for the Big Picture

Respiration is one of the most complex processes you will encounter in a beginning biology course, and the brief overview here hardly scratches the surface. Bear in mind, as you make this foray into biochemistry, that respiration is nothing more or less than the process by which cells obtain energy. It is the act of transformation—from food molecules into energy—that drives life itself.

Cellular Division

Cells form the structure of organisms and are the "factories" in which all the organism's life processes ultimately are initiated and maintained. Cells are also capable of another important activity. They divide, producing two cells where there was one. If your course includes a lab section, you will probably witness such cell division "live." It is one of the most fascinating of biological phenomena.

Cell division occurs when the cell reaches a certain size. Both prokaryotic cells (those "primitive" cells that lack an organized nucleus) and eukaryotes (which have a nucleus) replicate by division. We'll focus briefly on the process in eukaryotes.

Mitosis and Cytokinesis

Cell division in eukaryotes involves two main processes. In the first process, *mitosis*, the nucleus divides and the chromosomes, which hold the genetic material of the cell within the nucleus, are arranged and distributed to the daughter cell (as the new cell is called). In lab or on video, you will observe four major phases of mitosis: prophase, metaphase, anaphase, and telophase. These steps mark an orderly progression of chromosome formation from chromotin material, dissolution of the membrane and nucleolus of the nucleus, division of the chromosomes, and reformation of *two* new nuclear membranes around two new nuclei, each of which bears a newly formed nucleolus. In mitosis, the number of chromosomes is preserved without change in the daughter cells. (We will look more closely at the role of chromosomes in Chapter 15.)

Once nuclear division is completed in telophase, *cytokinesis*, the separation of the cytoplasm, begins. In animal cells (and other cells lacking rigid walls), the cytoplasm constricts in the middle, forming a dumbbell shape. The middle constricts until two independent cells—the daughter cells of the original cell—are formed.

In plant cells that have rigid walls, the cell does not assume a constricting dumbbell shape, but, rather, a *cell plate* consisting of bubble-like vesicles forms along the cell's equator.

You will encounter another phase in the *cell cycle*. *Interphase* is the portion of a cell's existence when it is not dividing. This is when most of the cell's growth and metabolism occur. Biologists divide interphase into three subphases:

- **G_1 (first gap phase), the period of rapid growth after cytokinesis; many of the cell's organelles appear during this phase**
- **S (synthesis phase), during which DNA synthesis occurs**
- **G_2 (second gap phase), which sees further growth and the double stranding of chromosomes preparatory to mitosis**

Meiosis

Mitosis is the process by which *somatic cells*—the organism's nonsexual cells—divide. Sex cells, also called *gametes*, divide by a different process, *meiosis*. Whereas mitosis produces daughter cells that preserve the original number of chromosomes (called the *diploid* number), meiosis produces *haploid* gametes—cells containing half the number of original chromosomes. When male (sperm) and female (egg) gametes unite (in fertilization), each carrying a haploid number of chromosomes, the original diploid number is restored.

Meiosis is a more complex process than mitosis and occurs in two main phases, which are each divided into their own subphases.

Meiosis I is the phase in which the reduction of the number of chromosomes, from diploid to haploid, occurs. The process consists of four phases, prophase I, metaphase I, anaphase I, and telophase I, followed by interkinesis, a brief phase in which the still undivided gamete cell has two haploid nuclei.

From interkinesis, the process continues through meiosis II, which is, mechanically, much like mitosis. Proceeding through prophase II, metaphase II, anaphase II, and telophase II, the result is ultimately division of the cell into *four* cells, each with a haploid nucleus bearing one member of each pair of chromosomes present at the start of meiosis.

Meiosis is a critically important process. The number of chromosomes within cells of a given species is constant. Normal human somatic cells, for example, have 46 chromosomes, and gamete cells have 23. A somatic cell from a cat has 38 chromosomes—19 in a gamete cell. By reducing the diploid to the haploid number, meiosis keeps the number of species chromosomes constant. If gametes each contained the full diploid number, the species number of chromosomes would be doubled, and the mechanism of heredity would not function. In fertilization, by combining two haploid cells from two parent organisms, it is made most likely that the offspring organism will inherit some traits from both parents while ensuring that the number of chromosomes characteristic of the species remains constant.

CHAPTER 9

TAXONOMY

AT SOME POINT EARLY IN MOST INTRODUCTORY GENERAL BIOLOGY COURSES, YOU will be introduced to the system of *binomial nomenclature* by which (with many revisions and modifications) biologists have been classifying living things since the Swedish botanist Carl von Linné (1707-1788)—better known by the Latin version of his name, Linnaeus—developed the system. Linnaeus based his classification of organisms on their morphology (anatomy and structure), grouping related organisms into an overarching *genus* and a more specific *species.* Thus, Linnaeus assigned each organism two names. A human being, *Homo sapiens*, is a member of the genus *Homo* and the species *sapiens.* By convention, the genus name is capitalized, and the species name put in lowercase. The two names are properly used together; one speaks of *Homo sapiens*, not just *sapiens.*

The Species Concept

The modern concept of species is not based primarily on morphology, but on whether organisms can mate and produce fertile offspring. For example, a toy poodle and German shepherd, while they may look quite different from one another, are of the same species; they can breed, producing fertile offspring. In most cases, organisms of dissimilar species cannot produce any viable offspring; however, in some cases, animals of closely related species can produce *barren* offspring. The product of a female horse and a male donkey, for example, is a mule, an infertile hybrid. (A female donkey and male horse produce a hinny, also infertile.)

Phylogenetic Taxonomy

Genus and species are part of a hierarchical taxonomic system in which eight major categories are recognized. In descending order, these are

- Kingdom
- Phylum
- Subphylum
- Class
- Order
- Family
- Genus
- Species

All organisms can be classified in terms of this system (although biologists sometimes debate and revise the appropriate classification of particular organisms). A human being, for example, belongs to:

- The Kingdom Animalia (multicellular organisms requiring preformed organic material for food and motile at least at some stage of life)
- The Phylum Chordata (animals with a *notochord*—an embryonic skeletal rod—a hollow nerve cord, and gills in the pharynx at some time during the life cycle)
- The Subphylum Vertebrata (organisms with a backbone enclosing a spinal cord, and a brain encased in a skull)
- The Class Mammalia (body with hair or fur at some time in life; female nourishes young with milk; lower jaw consists of a single bone)
- The Order Primates (arboreal mammals and their descendants; flattened fingers and nails—rather than claws; sharp vision, but relatively poor sense of smell)
- The Family Hominidae (bipedal; flat face with forward-looking eyes, binocular color vision; hands and feet specialized, for grasping and walking, respectively)
- The Genus Homo (long childhood, large brain, ability to use language)
- The Species *Homo sapiens* (reduced body hair, high forehead, prominent chin)

A careful reading of this classification may elicit the objection that human beings do not have gills, as chordates are supposed to have. A *more* careful reading, however, will note that, for an organism to be classified as a member of the Phylum Chordata, gills must be present *at some time in its life cycle.* Human fetuses do show gill structures, which disappear as the fetus develops. Similarly, while it is true that human beings do not live in trees (unlike most primates), they are believed to be descended from apelike creatures who were arboreal.

How Many Kingdoms?

For many years, biologists recognized five great kingdoms of organisms:

1. **Monera:** Single-celled organisms without an organized nucleus or membrane-bound organelles. Bacteria and blue-green algae were traditionally assigned this kingdom.
2. **Protista:** Single-celled organisms with an organized nucleus and other organelles. The amoeba is one example of a protist.
3. **Fungi:** Nonmotile, plantlike organisms that absorb their food from living or nonliving sources. Examples include mushrooms and various molds.
4. **Plantae:** The plant kingdom, including all plants other than fungi.
5. **Animalia:** The animal kingdom, including all multicellular animals.

Today, many biologists recognize six kingdoms. Instead of Monera, they assign bacteria to two kingdoms: Archaeobacteria and Eubacteria. See the next chapter for a discussion of these two kingdoms.

Purpose of Taxonomy

Taxonomy not only makes the study of living things easier by sorting them in a logical fashion, it also helps biologists more readily observe relationships among organisms. That is, one can study not just how one organism relates to another, but how an entire species, genus, family, and so on relates to other species, genera, families, and so on, and to the physical environment. Far from segmenting the view of the natural world, taxonomy tends to unify it, allowing for the emergence of a more holistic picture of life.

CHAPTER 10

BACTERIA, VIRUSES, VIROIDS, AND PRIONS

IT IS A LOGICAL STEP FROM THE STUDY OF THE ANATOMY AND PHYSIOLOGY OF THE CELL to the study of the most basic forms of single-celled life, the bacteria. And it is a logical step *backward* from bacteria to the even simpler viruses, which are not, properly speaking, living organisms at all.

The Prokaryote World

Bacteria (singular, *bacterium*) are single-celled organisms of the prokaryote type; that is, they lack an organized nucleus bound by a membrane. They also lack other membrane-bound organelles. Most modern biologists recognize two large bacteria kingdoms: *Archaeobacteria* and *Eubacteria*, or "true bacteria." While the two groups have important chemical differences, they both share the general characteristics of prokaryotes.

While murein makes the walls of bacterial cells strong, it can also figure as the bacteria's Achilles' heel. The antibiotic penicillin acts against bacteria by inhibiting the synthesis of murein. Since the cell walls of nonbacterial (eukaryotic) cells do not contain murein, penicillin can destory bacterial cells without harming body cells.

With the exception of *mycoplasmas* (the smallest living cells), bacterial cells are enclosed in cell walls made of *murein*, a substance consisting of polysaccharide molecules bound together by amino acids. The result is a cell wall of considerable strength.

The cell wall envelops a cell membrane. In some prokaryotes, this membrane is convoluted and serves as the site of the electron transport and enzyme processes necessary for respiration. Also attached to the membrane are *mesosomes*, to which the cell's DNA molecule is attached. During cell division, the mesosomes apparently function to separate the DNA chromosomes.

In contrast to eukaryote cytoplasm, that of the prokaryotes lacks organelles and other fine structures. Ribosomes, which appear as so many granules, are the only cellular organelles within the cytoplasm. These are the sites of protein synthesis.

The cell's genetic material consists of a single circular molecule of DNA. Bacteria replicate through a process called *binary fission*, in which the DNA chromosome (which appears in the cell cytoplasm, attached at the ends to mesosomes) is replicated before division. Next, the cell wall and plasma membrane grow inward, dividing the cell in two.

In cyanobacteria (blue-green algae), a peripheral membrane is also present in the cell. This contains chlorophyll and other pigments.

Prokaryotic cells (measuring from 0.5 to 10 micrometers) are smaller than eukaryotic cells. Some prokaryotic cells are capable of secreting a slimy capsule that protects the cell from destruction by white blood cells. Many bacteria are capable of locomotion, propelled by a whiplike structure called a flagellum.

The Archaeobacteria

The most distinctive feature of this relatively small group of bacteria is that they are *anaerobic*; that is, they cannot survive in an oxygen environment and live instead in the harshest environments earth offers: in stagnant marshes, in salt-water lakes and ponds, in the sludge at the bottoms of lakes and ponds, and in other places where anaerobic (airless) conditions prevail. Archaeobacteria exhibit a chemistry that sets them apart from all other organisms. Their cell membranes are made up of branched lipid molecules, in contrast to the straight-chain fatty acid molecules that make up the membranes of other cells. The archaeobacterial metabolic processes are also unique, because they use different enzymes than other cells employ.

Major groups of archaeobacteria include:

- The methanogens: By far the largest group of archaeobacteria, methanogens use hydrogen gas to reduce carbon dioxide, producing energy and, as a by-product, methane gas ("swamp gas").
- Members of the genus *Halobacterium* are "salt-loving bacteria," which live in salty environments and form energy-storing ATP molecules through a primitive form of photosynthesis.
- Thermophiles live in hot springs where temperatures approach the boiling point of water.
- Thermoacidophiles live not only in near-boiling environments, but in highly acidic ones.

The Eubacteria

These bacteria typically have rigid cell walls and may or may not be capable of motility (movement from place to place). They are often classified by shape:

- *Bacilli* are rod-shaped.
- *Cocci* are round.
- *Spirilla* are spiral-shaped.
- *Diplococci* are round bacteria that occur in pairs.
- *Streptococci* are round bacteria that occur in chains.
- *Staphylococci* are round bacteria that occur in clusters.

As mentioned earlier, reproduction in eubacteria is by binary fission: chromosomes are replicated and then the cell divides. This is much simpler—and faster—than mitosis. Bacteria, unchecked by environmental factors, multiply at an astounding rate.

A relatively small number of eubacteria are *autotrophs*, capable of transforming inorganic materials into organic compounds as food. Photosynthetic bacteria use light energy to accomplish this, while chemosynthetic bacteria can, without the presence of light, oxidize ammonia, nitrate, sulfur, or hydrogen compounds to produce high-energy organic compounds.

Most eubacteria are not autotrophs, but *heterotrophs*, which cannot synthesize their own nutrients but must obtain food from outside. Those that absorb nutrients from dead organic matter are *saprophytes* or *saprobes*, while those that live inside other organisms are *parasites.*

The four most important eubacteria groups are listed below.

1. **Myxobacteria** (slime bacteria) are saprobes that develop in a colony and move together as a mass of slime.
2. **Spirochetes** are anaerobic eubacteria and are highly motile, propelling themselves by a corkscrew motion. The best-known example of a spirochete is *Treponema pallidum*, which causes syphilis. (While many spirochetes produce disease, most bacteria are either harmless or beneficial—for example, they may break down harmful substances.)
3. **Rickettsiae** are nonmotile parasites that live in the cells of ticks and mites, whose bite transmits them to animals and humans. Members of this group produce such diseases as Rocky Mountain spotted fever and typhus.
4. **Actinomycetes** combine certain characteristics of eubacteria and fungi; like fungi, they are structured as multicellular filaments. They are saprobes, and most are harmless, although *Mycobacterium tuberculosis*, a member of this group, causes tuberculosis, and *Mycobacterium leprae* causes leprosy. Other members of this group produce substances that are the sources of a

variety of antibiotic medicines.

Cyanobacteria

This group is variously classified as blue-green algae or as bacteria. They are autotrophs, deriving their nutrients by photosynthesis. Cyanobacteria play a key role in the environment, producing oxygen as a by-product of photosynthesis. This is vital in replenishing the atmosphere. Cyanobacteria are also a food source for various vertebrate and invertebrate species living in lakes and oceans.

Viruses

Viruses are not related to bacteria and, indeed, are not living organisms. They lack a cell's apparatus for metabolism, and the only "lifelike" function of which they are capable is reproduction (replication). Nevertheless, because viruses, like bacteria, are so often associated with disease, they are frequently considered along with bacteria in introductory biology courses.

The virus particle, or *viron*, consists of a protein coat (*capsid*) surrounding a DNA or RNA core or a core consisting of both DNA and RNA. The shape of the virus, which is much smaller than the smallest cell, may be helical or polyhedral (many-sided).

A virus cannot exist independently, but must invade a "host" cell. After entering the cell, the nucleic acid of the virus "captures" the nucleic acid of the host, which it uses to make more virus nucleic acid and virus proteins. More viruses are thus replicated within the host cell until no more can fit. At this point, the host cell bursts open, releasing the viruses, which then invade more cells. Obviously, viral invasion is destructive to the host organism, and viruses are responsible for many diseases.

Viroids and Prions

Viroids, infectious particles even smaller than viruses, were discovered in the 1960s as the agents of certain plant diseases. A viroid consists only of a circular RNA molecule and lacks the protein capsid of a virus. Like the virus, the viroid is not a living organism.

Even stranger than the viroid is another class of disease-producing agent, the *prion*. An aberrant form of a normally harmless protein found on the surface brain cells of mammals and birds, the prion may enter the brain through infection or may arise from a mutation in the gene that encodes the protein. Once it is present in the brain, the prion multiplies by inducing benign (normal) protein molecules to refold into its own aberrant shape. Why this occurs is not yet understood; however, prions appear to be responsible for a number of fatal neurodegenerative diseases (collectively referred to as *transmissible spongiform encephalopathies*), including bovine spongiform encephalopathy, commonly called mad cow disease.

CHAPTER 11

PROTISTS AND FUNGI

THE PROTISTS ARE EUKARYOTIC, SINGLE-CELLED ORGANISMS WITH MEMBRANE-bound, organized nuclei. Fungi are also eukaryotic, but (except for yeasts) are multicellular. However, fungi cells differ from those of other species because they have no boundaries—or only partial boundaries—between cells, and cells have more than one nucleus.

The Protists

Biologists generally recognize three major protist groups: protozoa (animallike protists), funguslike protists, and plantlike protists.

Protozoa

The very word *protozoa* means "first animals." These organisms live in all sorts of environments, including fresh water, salt water, and soil. Some protozoans are parasitical and live inside other organisms. Four protozoan phyla exist:

1. **Mastigophora.** These organisms have one or more flagella, but not all species are motile. Some are *sessile*, attaching themselves to submerged surfaces and using their flagella to help take in food. Among this phylum is the genus *Trypanasoma*, whose members are parasites that cause sleeping sickness in human beings.
2. **Sarcodina.** The members of this phylum are amoeboid; that is, they move by means of *pseudopods*, free-flowing extensions of an amorphous body.

Pseudopods not only allow the organism to move, they also serve to capture food, which is stored and digested in food vacuoles within the cytoplasm. Another kind of vacuole, the *contractile vacuole*, maintains the water balance between the cell and its watery environment by continually expelling excess water from the cell.

Marine groups of Sarcodina, such as Radiolaria and Foraminifera, secrete hard shells around their amoeboid bodies. Over millions of years, these shells, the remains of dead organisms, have come to constitute much of the bottom mud of the ocean floor.

3. **Sporozoa.** These are the parasitic spore formers, whose lives are typically characterized by rather complex cycles that may involve multiple hosts (as in the case of *Plasmodium vivax*, the protozoan that causes malaria). For at least part of its life cycle (the mature stage), Sporozoa are not motile.
4. **Ciliata.** The most numerous and varied of the protozoa species, Ciliata organisms are all characterized by the presence of *cilia*, short, hairlike strands of cytoplasm used both for locomotion and for sweeping food into a cell opening called the *oral groove.* The members of this group typically have distinctive shapes, maintained by a rigid pellicle lying just inside the cell membrane. While most protozoans reproduce by asexual mitosis, some Ciliata also exhibit conjugation, a form of sexual reproduction in which two organisms join at the oral groove, whereupon a structure called the *micronucleus* of each undergoes meiosis as the nucleus of each disintegrates. One micronucleus from each cell migrates to the other cell. The haploid micronucleus in each cell fuses with a haploid gamete in each cell, and the joined cells now separate. A new diploid nucleus is formed in each cell. It divides, producing new organisms.

Funguslike Protists

The protists of this group behave like fungi in that they absorb their nutrients from dead organic matter or from the body of a living host.

Protomycota live as parasites or saprophytes and produce motile zoospores, some of which fuse with others in a form of sexual reproduction, and some of which develop asexually into a mature vegetative cell, which gives rise to more zoospores.

Gymnocota are slime molds and consist of two large groups, Myxomycota (true slime molds) and Acrasiomycota (cellular slime molds). True slime molds grow as shapeless masses of slime in damp soil or on rotting vegetative material. They exhibit a life cycle that begins with a multinucleate mass called a *plasmodium*, which is capable of amoeboid motility. After the plasmodium engulfs food in the form of bacteria and other organic matter, it stops moving and becomes a mass of *fruiting bodies*, which support *sporangia* (singular, *sporangium*), structures that produce spores. The spores undergo meiosis, producing flagellated gametes, which fuse to form a *zygote*, an amoeboid organism that gathers nutrients and undergoes mitosis. However, the mitosis produces multiple nuclei without cytoplasmic division, so that the multinucleate but undivided plasmodium is produced—and the cycle begins anew.

Cellular slime molds, Acrasiomycota, also have a distinctive life cycle, but one that is quite distinct from that of Myxomycota. They, too, produce fruiting bodies (called *sorocarps*), which produce spores that are carried by wind, water, or insects, and so on to new environments. Spores that find favorable environments germinate, "hatching" a *haploid* amoeba, which lives in the soil, consumes nutrients, and grows, until it undergoes mitosis to produce *haploid* daughter cells. These cells mass together into a *pseudoplasmodium*, sometimes called a slug. Although this mass is made up of individual organisms, it moves as one. At some point, the mass stops moving, and the constituent cells are differentiated into *stalk cells*, which give rise to fruiting bodies. Thus the life cycle begins anew.

Plantlike Protists

These organisms are especially interesting because they have characteristics of animals as well as plants.

The **Euglenophyta** are organisms that contain chlorophyll and carry on photosynthesis. Propelled by a flagellum, the organism can swim toward light, which it senses by means of an eyespot (the *stigma*).

The **Chrysophyta** include yellow-green and golden-brown algae as well as diatoms. Containing chlorophyll, Chrysophyta perform photosynthesis. They live in salt and fresh water, as well as in moist places between rocks. The diatoms differ from the algae in that their cell walls are impregnated with hard silica and pectin shells. These shells settle to the bottom of the water as diatomaceous earth, a substance valued as a filtration material.

The **Pyrrophyta** are propelled by *two* flagella of unequal length and so are called *dinoflagellates*. The paramecium, which you will doubtless examine in the laboratory, is the most familiar of the group. Some species (such as Noctiluca) develop the highly specialized property of bioluminescence: they give off light.

The Fungi

Fungi may be the first multicellular organisms you study in an introductory course (though yeasts, members of the same kingdom, are unicellular); however, the cells of fungi differ from those of other multicellular organisms in that boundaries between cells are either absent or only partially formed. Moreover, fungi cells are also *coenocytic*: they have many nuclei within a single mass of cytoplasm.

Fungi are heterotrophs—acquiring nutrients from outside of themselves—but they neither engulf their food (in the manner of an amoeba, say) nor ingest it through a mouth. Instead, they absorb it, either from dead organic matter or parasitically, from living organisms. This is accomplished by means of digestive enzymes secreted onto the food source. The function of these enzymes is to make the molecules of the food source small enough so that they can be absorbed by the fungus.

All fungi begin life as spores. The *spore* is a single cell that is protected by a tough coat and is capable of surviving in harsh and unfavorable environments. Spores are also quite transportable. Typically, they are carried by the wind, animals, or other means, and, if they settle in a favorable environment, they will germinate: the cell wall of the spore ruptures, the cell absorbs food, and a threadlike structure called a *hypha* develops. As the fungus grows, a tangled mass of hyphae develop into the body of the mature fungus, called a *mycelium.* Fungi generate a chemical that causes potentially competing hyphae to grow away from it. Thus competition for food is eliminated—a valuable property in an organism that cannot move to another food supply.

While most fungi reproduce primarily by means of spores (*sporulation*), some also reproduce vegetatively: new organisms are produced from a part of the parent's body. These processes of asexual reproduction are typically augmented by cycles of sexual reproduction as well. The fungus develops specialized *gametangia*, which produce gametes that fuse to form a zygote.

Fungi Types

Unlike the kingdom Protista, which is subdivided into a number of phyla, the kingdom Fungi has only a single phylum, Mycota, which encompasses five classes:

1. **Oomycota** are mostly water molds (a few land species also exist). The spores of this group are propelled by flagella. Some water molds are saprobes (living off dead matter), but a few are parasitic.
2. **Ascomycota** are the sac fungi, the most familiar examples of which are yeast, powdery mildews, blue and blue-green molds, truffles, and morels. In all, this class comprises some 30,000 species. Their name, sac fungi, derives from the *ascus* (little sac) in which reproductive cells are held during a portion of their life cycle.
3. **Zygomycota** includes the familiar black bread mold, *Rhizopus stolonifer.* The coenocytic cells of this organism's hyphae have walls composed of *chitin*, a tough polysaccharide that is also the principal substance in the exoskeletons of insects. Chitin is present in the cell walls of most fungi.

4. **Basidiomycota** are also called club fungi and include the many species of mushrooms, as well as bracket fungi and smuts. The species of this class are characterized by a *basidium*, or club, at the tip of each reproductive hypha. Within this basidium, two haploid nuclei fuse to form a diploid nucleus, which then undergoes meiosis, producing four haploid nuclei. Around each of these nuclei, a delicate, stalked *basidiospore* forms, which eventually breaks away in the wind, carrying with it spores.
5. The final fungus class is **Deuteromycota**, the "imperfect fungi." They earn this name because they do not exhibit life cycles typical of other fungi. Sexual reproduction plays no part in their reproductive cycles. Familiar imperfect fungi include the organisms that cause athlete's foot and ringworm.

The Lowly Fungi?

Some beginning students tend to look down on the fungi, and, indeed, while most fungi do not cause disease (though some do), many, such as bread mold, are annoying. Yet they play a vitally important role in maintaining ecological balance by breaking down the dead bodies of plants and animals, as well as animal waste matter. In the form of mushrooms, truffles, and morels, fungi are a significant food source, and the role of yeast in many food-preparation processes is, of course, extremely important. Fungi also help ripen cheese, and one imperfect fungus, *Penicillium notatum*, is the source of the breakthrough antibiotic, penicillin.

CHAPTER 12

THE PHOTOSYNTHETIC PLANTS

THE VAST KINGDOM PLANTAE IS COMPOSED OF ALL THE GREEN PLANTS, FROM MICROscopic unicellular organisms to towering redwoods. They are united by one great property: photosynthesis. The presence of the green pigment chlorophyll in the *chloroplasts* of green leaves and stems traps light, which becomes the energy that drives a chemical process that yields the chemical energy essential to the life processes of the organism.

Photosynthesis

Photosynthesis is a complex process that is fundamental not only to the life of green plants, but to all life, because it helps to replenish atmospheric oxygen by converting carbon dioxide into oxygen and because it sustains the *producers*, the green plants that are the base of the food pyramid—the ultimate source of food energy for the *consumers*, the animals at higher levels of the pyramid (see Chapter 18). The overall process may be expressed in a simple equation:

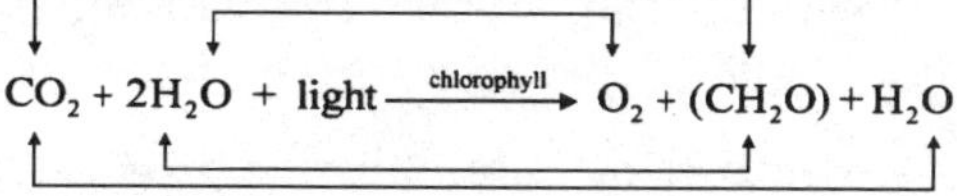

$$CO_2 + 2H_2O + \text{light} \xrightarrow{\text{chlorophyll}} O_2 + (CH_2O) + H_2O$$

The principal chemical reactions involved in photosynthesis are reduction-oxidation, or *redox*, reactions. *Oxidation* is the removal of an electron from a molecule, which results in the release of energy. *Reduction* is the addition of an electron to a molecule, which results in the storage of energy. When one substance is oxidized, another is reduced. Thus photosynthesis is a cycle of storage and liberation of energy. The CH_2O is the substance that is synthesized from sunlight and that is transformed (by the *Calvin cycle*) into glucose, which is used as the organism's food.

The chemical reactions of photosynthesis take place in the chloroplasts, the cellular structures that contain chlorophyll within sacs called *thylakoids.* Four major events occur during photosynthesis:

1. Photochemical reactions
2. Electron transport
3. Chemiosmosis
4. Calvin cycle

The photochemical reactions and electron transportation are carried out within the thylakoid membranes. These membranes surround a vacuole in which hydrogen ions are stored, which are used in the Calvin cycle, the process of carbon fixation by which the product of the photochemical reactions, CH_2O, is transformed into ribulose diphosphate (RDP) and, ultimately, into glucose, a high-energy 6-carbon sugar.

The Lower Plants

The plant kingdom may be very broadly divided into the lower and higher plants. In contrast to the higher plants, the lower plants are not differentiated into the plant structures most familiar to us: roots, stems, and leaves. Lower plants consist either of a single cell or an undifferentiated tissue of multiple cells. The lower plants comprise three kinds of algae.

Chlorophyta

The green algae, Clorophyta, include:

- various species of single-celled, chlorophyll-bearing organisms, some of which are motile (such as chlamydomonas)
- colonial forms—collections of unicellular organisms that, while independent cells, live in colonies (members of the genus *Volvox* are common examples)
- multicellular organisms without specialized cells (*Ulva,* or sea lettuce, is an example)

Members of Chlorophyta reproduce through various combinations of asexual and sexual means. Sea lettuce exhibits a particularly complex and interesting life cycle called *alternation of generations* because one generation of the organism is produced sexually by gametes and the next generation is produced asexually from zoospores.

Phaeophyta

These are the brown algae, which we know more commonly as seaweed. Although the root-stem-leaf differentiation of higher plants is absent from Phaeophyta, these organisms are the largest of the algae (some reaching lengths as great as a hundred feet), and they do exhibit cellular differentiation, including such features as *holdfasts* (which anchor the plants to rocks) and *air bladders* (which provide buoyancy).

Reproduction in Phaeophyta is sexual. In some species, the male and female sex organs are present on the same plant, while in others they are on separate plants.

Rhodophyta

These are red seaweeds, found in warm-water oceans, typically at depths well below those occupied by Phaeophyta. Note that while the red algae often exhibit feathery leaflike structures, these are not true leaves; like other lower plants, Rhodophyta is not differentiated into root, stem, and leaf. The life cycle of most red algae features alternation of generations.

The Higher Plants

Most of the plants that are most familiar to us belong to this division of the plant kingdom. They are the land plants, which have developed methods for conserving water and for sexual reproduction by means of special sex organs protected by non-reproductive cells. Typically, higher plant structures are specialized as roots, stems, and leaves. A vascular system carries and distributes water and other nutrients throughout the plant's body.

The Root

In higher plants, roots function to anchor the plant to the soil and to absorb water as well as dissolved minerals from the soil for use by the plant. The typical root exhibits a fairly complex cross-sectional structure.

- The outer layer of the root, the *epidermis*, not only protects the root, but is specialized in the absorption of water.
- The epidermis covers the *cortex*, which stores food and water.
- At the next level is the *pericycle*, from which branch roots are produced.
- The *xylem* conducts water and dissolved nutrients up through the root system into the stem.
- *Phloem*, at the same level as the xylem, conducts water and dissolved material down from the leaves, through the stem, and into the root.
- The innermost structure of the root cross section is the *parenchyma*, which stores food and water and functions to support the other root tissues.

The Stem

Stems conduct water and dissolved materials up from the root system and down from the leaves. Structurally, the stem produces and supports leaves and flowers. The stem also functions as a storage system for food.

Some stems are further specialized. The stems of vinous plants often have tendrils that allow the plant to climb walls and other surfaces. Some stems bristle with thorns, which defend the plant against predators. From the stems of some plants runners grow, from which new plants are propagated. *Rhizomes* are the specialized underground stems of ferns, which serve to reproduce plants. *Tubers* and *corms* are stem adaptations that specialize in the storage of food. Furthermore, some stems are *herbaceous*, soft and green, while some are *woody* and covered by a protective bark.

The Leaf

The leaves of the higher plants offer a substantial surface for exposure to light and moisture. Within the structure of the leaf, cells containing chloroplasts perform photosynthesis. The leaf is structured with special pores that allow for the passage of oxygen and carbon dioxide.

Bryophytes

The simplest, most primitive of the higher plants are the Bryophyta: mosses, liverworts, and hornworts. While these plants are multicellular, they lack true stems, leaves, and roots—though they have rootlike structures called *rhizoids*, which not only anchor the plant, but absorb soil moisture. Alternation of generations occurs.

Tracheophytes

These plants are well-adapted land dwellers, with highly developed vascular systems for conducting water up from the roots (through xylem) and down from the leaves (through phloem).

Psilopsida

It is an open question as to whether psilopsids still exist or are extinct. The plants that are sometimes classified as psilopsids lack roots and often lack leaves, but so have branched stems.

Lycopsida

Some 900 species of club mosses belong to Lycopsida. They have vascular tissue, true roots, stems, and leaves. Sporangia are clustered on the leaves, imparting to these plants a clublike appearance.

Sphenopsida

This small group consists of some twenty-five species of horsetail. As with Lycopsida, vascular tissue, roots, stems, and leaves are present. The life cycle of the horsetail exhibits alternation of generations. The sporophyte generation is the conspicuous plant, with small, scalelike leaves along a stem, terminating in a cone. The sporophyte's haploid spores give rise to a smaller plant, called a *prothallus*, which is the gametophyte generation. The zygote that remains attached to the gametophyte produces the new sporophyte generation.

Pteropsida

The many familiar species of ferns belong to this group. Ferns have a vascular system and an underground rhizome, which grows horizontally (in species of temperate regions) and from which new plants grow. Fern leaves are composed of many leaflets, which not only carry out photosynthesis, but bear, on their undersides, *sori* (singular, *sorus*), spore cases. The ripe spores fall off the leaves and are carried by the wind to new habitats. Many botanists believe that fern plants represent a step in the evolution of seed plants.

Spermopsida

The seed plants may be divided into two broad groups, *gymnosperms* and *angiosperms*.

Gymnosperms are woody *cone-bearing* plants. They may have evolved directly from ferns. The cones are the site of seed production. In some plants, two types of cones exist—*megasporophylls* (seed cones) and *microsporophylls* (pollen cones). In such plants, pollen grains are carried from the pollen cones to the seed cones, fertilizing the *ovule*, which then germinates.

Angiosperms are the flowering plants. The site of reproduction is not a cone, but a flower, which encloses the male and female sex organs. The fertilized ovule causes the plant's ovary to enlarge. This is the fruit of the plant, in which seeds are formed. Each seed contains an embryonic plant and a store of food for it.

Angiosperms are the most recently and highly evolved form of plant life. They are found in varied environments throughout the world and are certainly the most familiar of all plants.

CHAPTER 13

SIMPLE AND COMPLEX INVERTEBRATES

MOST BIOLOGY SURVEYS PROCEED FROM A CONSIDERATION OF KINGDOM PLANTAE to Kingdom Animalia, usually starting with the simplest group, the invertebrates, animals lacking backbones. Most of us don't think first of invertebrates when we think of animals (perhaps vertebrate species such as dogs, cats, lions, and tigers leap first to mind), but, in fact, about 90 percent of animal species are invertebrate.

Distinctive Animal Characteristics

Before we survey the invertebrates, we may take note of some characteristics common to all animals.

Animal bodies are typically quite *symmetrical.* In animals with a more-or-less spherical shape, the symmetry is *spherical.* In animals (such as a starfish) that are wheellike in shape, the symmetry is *radial.* In animals with a recognizable anterior and posterior and left and right sides, symmetry is *bilateral.*

Animal tissue also develops in a distinctive way. All of the animal's structures develop from two or three embryonic cell layers called the *primary germ layers*: the *ectoderm* (outside skin) and *endoderm* (inside skin), and, in more complex animals, the *mesoderm* (middle skin) as well. Animals usually have an internal body cavity (*coelom*) lined with endoderm.

All members of the Kingdom Animalia are multicellular heterotrophs—dependent on external sources for food. Nutrients are used for energy as well as growth and tissue building. Energy not immediately used is stored as fat. Most (but not all) animals are capable of movement from place to place—though some water-dwelling invertebrates are sessile, anchoring themselves to a rock and using cilia or tentacles to sweep food (usually smaller organisms) into a digestive cavity.

Phylum Porifera

The phylum includes some 15,000 species of sponges, an organism composed primarily of undifferentiated cells arranged around a central body cavity. Support is provided by a "skeleton" made up either of needlelike crystals of lime or silica called *spicules* or a fibrous protein called *spongin.* Sponges are sessile animals who feed by filtering microorganisms from the surrounding water. *Collar cells,* which line the sponge's body cavity, have flagella, which create a current that draws the water and microorganisms in.

Sponges reproduce asexually by budding: a new organism grows from the parent's body, then splits off. Sponges also reproduce sexually. Motile sperm fertilize nonmotile egg cells. A single organism produces both egg and sperm. In some species, fertilization takes place within the sponge's body cavity, while in others the gametes are carried out to sea. In either case, the zygote develops into an embryo, which leaves the sponge and is briefly motile before attaching itself to a rock, on which it grows into a mature sponge.

In addition to reproduction, sponges can also regenerate. If a sponge is pulled apart, the cells will coalesce to form a new and complete sponge.

The Coelenterates

The sponges are the most primitive members of Kingdom Animalia. The next step up are the Coelenterates, members of Phylum Cnidaria. These include hydrozoa (members of Class Hydrozoa), jellyfish (Class Scyphozoa), and sea anemones (Class Anthozoa).

Coelenterates have bodies shaped like a hollow sac consisting of an ectoderm and endoderm. Two general body forms in this phylum are the *medusa* and the *polyp.* The medusa form is best illustrated by a jellyfish: an inverted bowl-shaped body, with top side convex and the bottom concave. The polyp form is cylindrical, and is readily illustrated by the hydra or by the sea anemone.

Coelenterates have a central gastrovascular cavity, in which digestion takes place. Food is taken in and waste expelled through the same opening in the cavity. Members of Phylum Cnidaria possess *cnidocytes,* stinging cells arrayed along tentacles, which secrete a toxin that paralyzes the organism's prey.

Platythelminthes

Planaria, a favorite laboratory animal in introductory biology courses, is one of the numerous flatworm species that make up this phylum. Planaria, like all flatworms, represent a significant step up the evolutionary ladder from Phylum Cnidaria. The animals exhibit differentiated excretory, nervous, and reproductive systems and exhibit bilateral symmetry with definite head and tail ends. Most flatworms (such as flukes and tapeworms) are parasites, with often complex life cycles involving development in various host organisms. A number of the parasitic flatworms cause disease.

Flatworms reproduce sexually, but are hermaphroditic, containing the sperm and egg within the same individual. Some planaria also reproduce by *parthenogenesis*, the offspring developing from unfertilized eggs. Planaria (and other flatworms) are also capable of regeneration if cut apart.

If you examine planaria in the laboratory, you will see an intestine and a central nerve cord. At the head end, the nerve cord ramifies as anterior *ganglia*, a group of nerve cells that may be taken as the evolutionary precursor of a brain. Eyespots (*ocelli*) at the head end distinguish between light and dark, and the head region also has *chemoreceptors* that aid the organism in locating food.

Nematoda

Roundworms are small, smooth, worms tapered at both ends. They possess well-developed, differentiated organs, including branching nerve cords terminating in anterior ganglia. They have eyespots. The mouth parts of this phylum are typically complex. Male and female sexes are present. Some roundworms are parasitic, including hookworms, which enter the host as embryos, burrowing into the host's feet and thence into the blood vessels. Transported in the bloodstream to the lung tissue, they bore through the bronchi and into the windpipe and throat. Swallowed, they pass into the intestine, where they attach themselves to the intestinal wall and live off the host's blood.

Annelida

The common earthworm—and other worms with segmented, round bodies—are members of this phylum. While a few are parasitic, most are free-living, occupying niches in the soil or in fresh or salt water.

The earthworm is often the first animal specimen dissected in introductory biology lab sections. Dissection reveals a highly developed system of organs, including a digestive system composed of a mouth, pharynx, esophagus, crop, gizzard, intestine, and anus. A closed circulatory system is also present, with blood contained in vessels and pumped by aortic arches—primitive hearts.

Some annelid species are hermaphroditic, and some *dioecious* (having separate sexes). While earthworms are examples of hermaphroditic annelids, they do not self-fertilize. At the *clitellum*, a band of smooth tissue on the outside of the earthworm's body, two earthworms copulate, exchanging sperm.

Mollusca

Mollusks—including clams and scallops (Class Pelecypoda), snails and slugs (Class Gastropoda), chitons (Class Amphineura), and octopuses and squid (Class Cephalopod)—make up this large phylum of about a thousand species. The mollusks share an unsegmented soft body type that is typically protected by a calcium carbonate shell secreted by a structure called the *mantle*.

Mollusks are an advanced invertebrate species, with very well developed organ systems, including a circulatory system in which blood is pumped by a two-chambered heart. The nervous system has three pairs of ganglia. The digestive system consists of a mouth, esophagus, stomach, digestive gland, intestine, and anus. Reproduction is usually sexual, and the phylum has two separate sexes.

Arthropoda

The largest phylum in Kingdom Animalia, Arthropoda is composed of some 800,000 species—more than all of the other animal species put together. Insects, spiders, ticks, mites, lobsters, crabs, millipedes, and centipedes are all members of the phylum, the name of which means "jointed foot"—a characteristic that unites members of this most diverse group.

In addition to jointed appendages, individuals of this phylum exhibit a *hemocoel*, or blood cavity, which is part of an *open* circulatory system. Instead of being supplied by closed blood vessels, organs are bathed in *hemolymph*. Arthropod respiration may be through *tracheal tubes* leading to pores called *spiracles*, or by means of gills.

While sexes are separate in Phylum Arthropoda, some insects reproduce by parthenogenesis, their eggs developing without fertilization.

The major classes of arthropods are Arachnida (spiders, ticks, mites, horseshoe crab, etc.), Crustacea (crabs, shrimps, lobsters, etc.), Diplopoda (millipedes), Chilopoda (centipedes), and Insecta. This last class is both the most advanced of the phylum and the largest. Although about half a million insect species have been identified, most biologists believe that unidentified species run into the millions. Insects are the dominant form of terrestrial life and exhibit a diversity unmatched by any other animal (or plant) class.

Their diversity notwithstanding, all insects share certain readily definable characteristics:

- The body is segmented, divided into head, thorax, and abdomen.
- The head bears a pair of compound eyes (multiple lenses) and one or more simple eyes (single lens), as well as antennae and mouthparts.
- Each of the three body segments that make up the thorax bears a pair of walking legs, a total of six legs in all.
- Attached to the last two thoracic segments (usually) are a pair of wings. These wings may be functional for flight or may be adapted as armor (as in beetles).
- The abdomen bears no appendages. Spiracles, the respiratory organs connected to the trachea, are present on the abdomen.

Insects also typically exhibit multistage life cycles in which the young organism does not resemble the adult. Most of us are familiar with the development of the butterfly from egg, through wingless caterpillar larva, through cocoon pupa, to winged adult. Such a life cycle is called *metamorphosis* and is an example of the operation of *hormones*, protein molecules secreted by endocrine gland cells, which trigger certain metabolic activity in *target* organs. Hormones regulate metamorphosis in insects.

The Importance of Invertebrates

Invertebrate species have profound effects on human life, both positive and negative. Perhaps the most far-reaching effects are created by the insects—not surprising, given their abundance. While some insects, such as honeybees (which provide honey and pollinate flowers) are obviously of great benefit to humankind, many other insect species are destructive. Think of the huge investment civilization routinely makes in substances and other means to control insect populations in an effort to avoid wide-ranging destruction of crops, and it becomes clear that the Order Insecta represents humankind's most formidable competition for food resources.

CHAPTER 14

THE CHORDATES

THE PHYLUM CHORDATA INCLUDES ALL ANIMALS THAT HAVE A *NOTOCHORD* AT SOME time in their life cycle. A notochord is not a backbone, but a living, rodlike structure extending the length of the body. In the lower—invertebrate—chordates, the notochord prevents the body from shortening when the muscles contract. In the higher—vertebrate—chordates, the notochord is present only during the embryonic stages of development and is replaced by the vertebrae, the backbone or spinal column, by the time the organism is ready to be born.

Other characteristics shared by the chordates include:

- Embryonic development from three germ tissue layers
- Bilateral symmetry, with anterior-posterior differentiation
- True digestive tract, beginning with a mouth and ending with an anus
- Presence of pharyngeal gill slits at some stage of development
- Presence of dorsal hollow nerve cord (*neurocoel*)

Echinodermata

Closely related to Phylum Chordata is Phylum Echinodermata, which represents the evolutionary step leading up to the chordates. Animals in this phylum don't look like the more advanced prevertebrate chordates, let alone the vertebrates. They are spiny-skinned sea animals, including the sea lily (Class Crinoidea); the sea star (Class Asteroidea); the brittle star and serpent star (Class Ophiuroidea); the sand dollar, sea biscuit, and sea urchin (Class Echinoidea); and the sea cucumber (Class

Holothuriodea). Members of Class Crinoidea are sessile, while those of the other classes are free-moving. The echinoderms are characterized by:

- Spiny skin (the meaning of the word *echinoderm*)
- Pentaradial symmetry; echinoderm bodies have five *antimeres* radiating from a central disc with the mouth in the middle.
- Complete digestive system, but with nonfunctional anus
- A vascular system that, by regulating pressure in tubes, moves the animal
- Absence of a head
- Absence of an excretory system
- Absence of a respiratory system

Pentaradial symmetry is, of course, quite different from the bilateral symmetry that members of Phylum Chordata share; however, the echinoderms do exhibit bilateral symmetry in their larval stages.

Hemichordata

This phylum is an intermediate evolutionary link between the echinoderms and the chordates. The hemichordates are worm-shaped inhabitants of shallow seas. Their conical *proboscis* resembles an acorn—hence their common name, acorn worms. They possess a rudimentary open vascular system that pumps colorless blood to the animal's body wall and gut.

Tunicata and Cephalochordata

Phylum Chordata proper encompasses two prevertebrate subphyla, Tunicata (also called Urochordata) and Cephalochordata. Tunicata get their collective name from the tough "tunic" of cellulose these marine animals (commonly called sea squirts) "wear." Cellulose is a material not normally found in animals. Some Tunicates are bottom-dwelling sessile filter feeders, while others swim freely.

The cephalochordates are commonly called lancelets and are some twenty species of small, fishlike marine invertebrates. These organisms are the closest living relatives of the vertebrates, with bilateral symmetry, a head, tail, and well-developed nerve cord and notochord.

The Vertebrates

The most advanced of the chordate subphyla, Vertebrata consists of animals with a true segmented backbone surrounding a notochord, which the backbone may (during embryonic development) obliterate and replace. Vertebrates come in a great variety of forms, ranging from very small water and land animals to giant whales and elephants. Despite the tremendous variations among them, they share certain key characteristics in addition to the true backbone. These include:

- Clear development of a head with a brain enclosed in a cranium
- A closed circulatory system through which blood is pumped by a heart with at least two chambers, an atrium and ventricle
- Red blood with hemoglobin, a substance that carries oxygen to the animal's organs
- An hepatic portal system, which transports food-laden blood from the intestines to the liver, which processes the blood for distribution to body cells
- A post-anal tail, or vestigial remains of one, as in the coccyx ("tail bone") of human beings
- Never more than two sets of paired appendages (some vertebrates show only vestigial appendages)
- A mouth with movable lower jaw
- The presence of a thyroid gland

Agnatha

The organisms of this class are the most primitive of the vertebrates, the jawless fishes, including lampreys and hagfish. They are characterized by a round mouth and are without appendages. The notochord is present throughout the organism's life, and the skeleton consists of cartilage rather than bone. Lampreys are parasitic, attaching to fish or other organisms by means of its mouth, then breaking the organism's skin with its rasplike tongue. It then sucks the host's blood. Hagfish are scavengers, feeding on dead or dying fish.

Chondrichthyes

The best-known (*infamous* may be a better word) of this class are the sharks. Dogfish, rays, and skates are also members of Chondrichthyes. All are characterized by cartilaginous skeletons, but in contrast to Class Agnatha, they have paired appendages (fins), a lower jaw, and gill arches. The predatory ways of the shark also stand in stark contrast to the parasitic or scavenging feeding habits of Agnatha. Sharks are capable of great velocity and agility, and their sense organs are highly developed, including eyes (remarkably similar to the human eye) and olfactory receptors. Sharks also have *lateral line systems*, electrochemical receptors that aid in locating prey.

Bony Fish

Traditional biological classification puts all fish with bony (as opposed to cartilaginous) skeletons in Class Pisces. Many biologists, however, have abandoned this classification and now categorize bony fish into four distinct classes:

1. **Dipnoi.** Lungfish, organisms that use both gills and lungs for breathing. These animals live in seasonally dry estuaries. During the wet season, they live in the water and breathe through gills. During the dry season, they *estivate* (become dormant) in the mud, breathing through lungs. Lungfish have some body structures that resemble land-dwelling species. Many biologists believe the lungfish represent an evolutionary link between water-dwelling species and the amphibians (which they especially resemble early in their life cycle).
2. **Crossoterygii.** Most of these are extinct. They are the ancestral line from which modern fish as well as amphibia are thought to descend.
3. **Ganoidei.** The most primitive of bony fish, this group includes the sturgeon, gar, pike, and dogfish. Their most distinctive feature are their *ganoid* scales — a form of protective armor.
4. **Osteichthyes.** Most fish are members of this class of "true bony fish."

Amphibia

Frogs and toads (Order Anura) are the most familiar members of this class, which also includes Apoda (wormlike ground burrowers) and Urodela (salamanders, newts, and mud puppies—organisms that retain the tadpolelike tails that frogs exhibit only in early stages of development). Amphibians live in and out of the water (but return to the water to reproduce), and many species have limited ability to regenerate organs as well as the ability to estivate in order to survive unfavorable seasonal conditions.

Frogs exhibit a well-studied metamorphic life cycle, developing from waterborne external eggs as exclusively aquatic, legless, tail-bearing tadpoles, which gradually develop the legs and other body features of the mature frog, at which time the animal emerges from the water.

Reptilia

While the amphibians live on land as well as in the water, they must return to the water to bear their young. In contrast, most of the members of Class Reptilia are true land dwellers, who do not need to return to the water to reproduce.

Major reptile groups include many extinct forms, for which an abundance of paleontological evidence exists (fossilized remains), and four orders of living forms:

1. **Rhynchocephalia.** This group is represented by a single living species, the tuatara of New Zealand, a primitive lizard. Among its primitive characteristics are vertebrae with a fishlike structure.
2. **Chelonia.** These are turtles and tortoises, characterized by the distinctive hard, boxlike "shell," which is a development of the animal's skeleton.
3. **Squamata.** This order includes lizards and snakes. Both types of animals have bodies covered by many small, flexible scales, but whereas lizards have movable eyelids, visible ear pits, and (usually) legs, snakes are legless and also lack ear pits and eyelids. Despite the absence of legs, snakes use their many ribs to move quite effectively. Some lizard and snake species secrete poisonous venom with which they kill their prey, injecting them through syringelike fangs.
4. **Crocodilia.** Crocodiles and alligators belong to this order. The most highly developed of the reptiles, these animals have a two-chambered heart, the ventricle of which is *almost* completely divided in two as well. Their brains are relatively large. Amphibious, they are better adapted to movement in the water than on land, where their stubby legs move them quite slowly. These animals are aggressive predators whose powerful jaws are lined with a formidable array of teeth.

Aves

Birds are members of Class Aves. Most, but not all, are capable of flight; however, their truly distinctive characteristic is the presence of feathers—unique in the animal kingdom.

Feathers are one of the properties of birds that adapt them for flight. They are light and, by keeping the bird warm, they generate body heat, which warms the air surrounding the bird, causing that air to rise and thereby providing additional lift. Other adaptations for flight include:

- Sturdy but hollow, light bones.
- Aerodynamic wings.
- Highly developed eyesight.
- Well-developed brain.
- Four-chambered heart, permitting complete separation of oxygenated and deoxygenated blood for more efficient respiration. This enables the bird to maintain the high energy levels necessary for the muscular activity of flight.
- A well-developed digestive system, which includes a means of eliminating feces. Retaining feces would impair the ability to fly.

It is believed that the evolution of birds is directly linked to a reptilian ancestor, the extinct Archaeopteryx, whose feathers are an evolutionary development of reptilian scales, but who retained the teeth and bony tail of the reptiles. The scaly feet of birds still suggest a reptilian heritage—though the feet of birds have become specialized well beyond those of reptiles and are adapted to many uses. Another major evolutionary advance beyond the reptiles is *homiothermy*—"warm-bloodedness." Unlike "cold -blooded" reptiles, birds maintain a constant body temperature.

Among the most characteristic (and beautiful) adaptations of birds are coloration and song. Birds typically exhibit pronounced *sexual dimorphism*; that is, the sexes are clearly differentiated (in most cases) by the color of the feathers. Males are typically bright and showy, while females are more subdued, even drab in color. Bird song, produced by a *syrinx*, a secondary larynx, functions, among other things, to establish territorial rights. Indeed, birdsong is only one aspect of complex behavior that makes birds fascinating. Birds are monogamous (that is, one female mates with one male either for life or for the duration of the mating season), and, prior to mating, birds engage in a variety of often elaborate courtship behaviors. While fertilization takes place internally, the female lays eggs, which develop externally, usually in a characteristic nest built (or found) by one or both parents. The male and female take turns *setting*, sitting on the eggs in the nest to incubate them. Both parents typically feed and care for their young as well.

Mammalia

Members of Class Mammalia, the large class to which human beings belong, are the most advanced of the vertebrates and share the following characteristics:

- Females nourish their young with milk from mammary glands. The name *mammal* is derived from the name of these glands.
- All mammals have hair (at least at some point in their life cycle).
- All mammals are warm-blooded.
- Mammal red blood cells lose nuclei and mitochondria when mature; therefore, they serve exclusively to transport oxygen, using none themselves.
- Mammals have three basic kinds of specialized teeth: incisors, which tear; canines, adapted to biting; molars (and premolars), for grinding.
- Most mammals have seven distinct neck (cervical) vertebrae.
- The thoracic and abdominal cavities are separated from one another by a muscular diaphragm.

There are various ways to divide Mammalia. Most biologists recognize three subclasses:

1. **Protheria.** These are members of the Order Monotremata, the most primitive of mammals, including the platypus, the spiny anteater, and the long-snouted anteater. In contrast to other mammals, monotremes lay eggs externally.
2. **Metatheria.** Like the Subclass Protheria, Metatheria consists of but a single order, Marsupialia, of which the opossum and kangaroo are the most familiar examples. (Other fairly familiar examples are the koala, Tasmanian wolf, wombat, wallaby, and Australian native cat.) Marsupials lack a placenta, so that the young are born in a highly immature condition, able only to crawl into their mother's external "pouch" (*marsupium*). Here they are protected as they mature and are nourished on mother's milk.

   That most marsupials live exclusively in Australia suggests that the geographical isolation of the island continent permitted these primitive mammals to survive without competition from more advanced species.
3. **Eutheria.** This is the largest group of mammals, all of which have a *placenta.* The young of this subclass develop wholly within the mother, in the *uterus.* The placenta unites the fetus to the uterus, mediating metabolic exchanges of the fetus, serving the functions of nutrition, respiration, and excretion.

The range of placental mammals is vast. The most familiar and important orders include:

- **Cetacea.** Whales, porpoises, and dolphins
- **Insectivora.** Hedgehogs, moles, and shrews (all insect eaters)
- **Dermoptera.** Flying lemur
- **Chiroptera**. Bats (these are the only mammals capable of true flight)
- **Carnivora.** Dogs, wolves, lions, tigers, and so on (meat-eating predators)
- **Rodentia.** Hares, squirrels, rats, mice, beavers, and so on (plant-eating animals that gnaw)
- **Primates.** Lemurs, tarsiers, monkeys, gorillas, etc., as well as human beings

The most advanced of mammals, primates are characterized by *lack* of strong specialization in structure; that is, body structures are adaptable to a wide range of uses. Primate hands and feet are *prehensile* (grasping), usually with opposable thumbs and great toes. Nails are flattened rather than claws. Vision is acute and binocular. The brain is large, with a large cortical (surface) area created by cortical folding. The species' young require parental care for a long period of time.

CHAPTER 15

ANIMAL SYSTEMS: AN OVERVIEW

IN ADDITION TO STUDYING VARIOUS TYPES OF PLANTS AND ANIMALS IN A BIOLOGY SURvey course, you will also take a closer look at the principal animal *systems*. Recall from the discussion of cells in Chapter 8, that within a multicellular organism, functionally related cells are organized into groups called tissues, and functionally related tissues are, in turn, organized into organs. These structures are further organized into coordinated systems.

Digestion

Like any other organism, animals require nutrients, energy-rich organic compounds that enable the animal to carry on the processes of life. Animals are heterotrophs that require preformed organic molecules. These nutrient substances must be processed by the body, converted into forms that can be absorbed and metabolized by the body's cells. This processing is called *digestion*, and it is carried out by the digestive system.

It is likely that you will study the digestive system—and the other major animal systems—as they operate in human beings, which typically represents the most thoroughly developed form of any system.

The human digestive system begins with the *mouth*, which functions to take in food and break it up. The mechanical processing of food by the teeth is augmented by chemical processing as *saliva*, a special glandular secretion, both softens the food and begins the digestion of starch, breaking down the large starch molecules into smaller maltose molecules.

Food, in the form of a mass called a *bolus*, proceeds down the *throat* and into the *esophagus*, which pushes the bolus down into the stomach by means of muscular contractions (*peristalsis*) and gravity.

The stomach is a highly expandable muscular pouch that churns the bolus, mixing it with gastric juices into a soupy semifluid called *chyme*. The gastric juices are a combination of pepsinogen and hydrochloric acid. Together, these break down large protein molecules into smaller protein molecules called *peptides*.

From the stomach, the chyme enters the *small intestine*, a folded tubular structure some twenty-three feet long (in adults). *Vili*, fingerlike projections lining the walls of small intestine, digest and absorb the nutrients in the chyme, aided by enzymes secreted by the *pancreas* and by gall secreted by the *gall bladder* (which helps break down fats). Digestion takes place, for the most part, in the first section of the small intestine, called the *duodenum*, and absorption of the digested nutrients takes place in *jejunum* and *ileum*, farther along the small intestine.

Indigestible substances are passed along to the *large intestine* (also called the *colon*), which absorbs water from the material passed along by the small intestine and stores and eliminates the waste matter as *feces*.

Another organ, the *liver*, participates in digestion by converting excess blood glucose into glycogen, which is stored by the body as fat. The liver also converts certain amino acids into compounds that can be metabolized, and it processes fats in order to make these substances useful in metabolism. Finally, the liver stores vitamins and minerals, synthesizes certain blood proteins and cholesterol, and produces bile for the digestion of fat.

Respiratory System

Gas exchange is vital in any organism. In human beings, the *respiratory system* begins with the nose and *pharynx* (throat), where air is taken in and "conditioned"—warmed and moistened, as well as filtered by cilia in the nasal chambers. The air then passes down the windpipe (*trachea*), which branches into the *bronchi*, which in turn branch into *bronchioles*, at the end of which are tiny air sacs called *alveoli*. These structures lie within the *lungs*, which are composed of some 300 million alveoli. They are the sites of gas exchange. Red blood cells pass through thin, narrow capillaries in single file in the alveoli. Oxygen is transferred to them and binds with the hemoglobin contained within the red blood cells. The red blood cells also carry carbon dioxide waste produced by cellular respiration to the alveoli. This CO_2 enters the lungs and is expelled by exhalation.

The lungs are not muscular. They inflate and deflate passively as a result of the contraction and relaxation of the *rib muscles* and *diaphragm.* When these muscles contract, the volume of the chest cavity is increased, thereby lowering the air pressure within the cavity, so that air from the outside rushes in to fill the void, and the lungs expand. When the rib muscles and diaphragm relax, the chest cavity is constricted, forcing the alveoli to close while also expelling the CO_2 that is in the lungs.

While this process can be controlled, to some degree, by the will of the individual, breathing is largely an involuntary function of the *autonomic nervous system.* These involuntary nerve impulses are regulated by the level of CO_2 in the blood. A high level of CO_2 stimulates an increased rate of nerve impulses, which results in a higher breathing rate.

Circulatory System

Human beings have two major *circulatory systems,* one that circulates blood to the body tissues, and another that circulates *lymph,* which flushes out foreign particles from tissues and *phagocytizes* (destroys) them through filtration at the *lymph nodes.*

The central organ of blood circulation is the *heart,* a four-chambered muscular pumping organ. Oxygen-poor blood enters the chamber called the *right atrium* through a large vein called the *vena cava.* The blood passes through a valve into the *right ventricle,* from which it is pumped through the *pulmonary artery* to the lungs. Here gas exchange takes place: the red blood cells are oxygenated, and they give up their CO_2, which is expelled by the lungs upon exhalation. The oxygenated blood now returns to the heart—this time entering the *left atrium*—via the *pulmonary vein.* It flows from the left atrium through another heart valve (the *mitral valve*) and into the *left ventricle,* which sends it through the *aorta,* a large artery, from which it is distributed through a ramified system of arteries to the various tissues of the body.

The red blood cells in the blood contain *hemoglobin,* which loosely binds with oxygen and carbon dioxide, so is an ideal transport for both of these gases. At *capillaries,* tiny blood vessels throughout the body, the red blood cells give up their oxygen, thereby oxygenating the body cells, and pick up the waste product of cellular metabolism, CO_2. The input of the capillaries is the *arteries,* carrying oxygen-rich blood, and the output is the *veins,* carrying oxygen-poor and CO_2-laden blood back to the heart for pumping to the lungs.

In addition to the red blood cells, blood contains *leukocytes* or white blood cells, which phagocytize invading bacteria and viruses. The white blood cells play a critical role in the body's immune system.

In contrast to red and white cells, *platelets* are nonliving blood fragments, which are essential to blood clotting. Clotting is essential to controlling bleeding from wounds.

The useful lifespan of a red blood cell is about 120 days, after which it is destroyed by the *spleen*. New red blood cells (and white cells) are continually produced in the *bone marrow*.

Excretory System

A prime directive of any organism is the maintenance of *homeostasis*, a relatively stable internal environment. The ongoing processes of metabolism produce waste products, which, if not eliminated from the body, would destroy homeostasis and ultimately prove fatal.

The major organs of excretion in human beings are the two *kidneys* located on either side of the backbone at about the level of the stomach. Blood enters the kidneys through the *renal arteries*. It is distributed in the kidneys to *nephrons*, which filter the blood, reclaiming usable materials while collecting waste products (urea, uric acid, salts, etc.). These waste substances are formed into urine, which flows through vessels called *ureters* to the *urinary bladder*. When this structure fills, it becomes distended, stimulating excretion of urine through the *urethra* and out of the body. In the meantime, the blood, now filtered, exits the kidneys by way of the *renal veins*.

Skeletal System

Animal *skeletal systems* serve two major purposes: to provide support for the body and to aid in movement. Animals exhibit three basic skeletal types:

1. **Hydrostatic skeleton.** This is a water-based skeleton. The animal's muscles are supported not by bone, but by a fluid-filled body cavity. Worms are examples of animals that have a hydrostatic skeleton.
2. **Exoskeleton.** The bodies of animals such as mollusks and arthropods are externally supported.
3. **Endoskeleton.** Most of the higher animals, including human beings, have internal skeletons.

In animals with hydrostatic skeletons, movement occurs when muscles contract, squeezing the internal fluid against the skin, causing the worm to stiffen and the body to shorten and widen. When the contraction is relaxed, the animal's body expands again. Through repeated contraction and expansion, the animal moves.

In animals with exoskeletons, the exoskeleton typically provides a firm platform for muscle contraction.

In animals with endoskeletons, the relationship between muscles and bones is more complex. The bones of the skeleton are joined to one another through various joints and linked by *ligaments*. Muscles are attached to bones by *tendons*. The endoskeleton serves to transmit the force of muscle contractions, thereby creating movement.

Muscular System

The muscles attached to the tendons are composed of many thousands of *muscle fibers.* Each fiber is a specialized individual cell. Each cell contains *microfilaments*—thick ones composed of the protein *myosin,* and thin ones composed of another protein, *actin.* The microfilaments are arranged parallel to one another, such that they overlap, giving the muscle fiber a *striated* appearance.

When stimulated by a nerve impulse, the muscle contracts. The impulse enters the membrane of the muscle cell (the *sarcolemma*) and spreads across the cell, entering the cytoplasm (called, in muscle cells, the *sarcoplasm*). The impulse causes the actin filaments to slide across the myosin filaments, an action that pulls the ends of the muscle fiber in, creating a contraction. The sliding filaments require calcium ions and energy in the form of an ATP packet. *Cross-bridges* hold the filaments together during contraction, but after the contraction has taken place, the energy required to sustain the contraction is used up, the cross-bridges break, and the filaments slide back to their original position.

Muscle cell contraction is binary: either the filaments contract or they don't. The variability of contraction of entire muscles and muscle groups is a function of the *number* of fibers that contract, not the degree of contraction of individual fibers.

Striated muscle is only one type of human muscle. It is found in *skeletal muscle,* which is also called *voluntary muscle,* because it is under the control of the individual. Two types of *involuntary muscle* are also found in the human body. *Smooth muscle* has few actin and myosin filaments (so appears smooth). It controls autonomic or involuntary muscular actions, such as the constriction and dilation of blood vessels, the constriction of the urinary bladder, the peristalsis in the digestive tract, and so on. Another involuntary muscle, *cardiac muscle,* is striated (it contains many myosin and actin filaments), but is found only in the heart and is not under volitional control.

The arrangement and massing of muscles relative to the skeleton determine the mechanics of movement.

Nervous System

Muscles contract when they receive nerve impulses. The *nervous system,* an exclusive property of the animal kingdom, begins with specialized nerve cells called *neurons.* The human body has some 12 billion neurons. Each neuron consists of a cell body, from which one or more short extensions, *dendrites,* and one long extension, the *axon,* project. Axons are sheathed in a fatty envelope of material called *myelin.*

Bundles of myelin-sheathed axons function as nerves.

Animals have three types of neurons:

1. **Sensory neurons** receive stimuli from the external environment.
2. **Motor neurons** transmit impulses from the brain and spinal cord to muscles or glands, stimulating contraction or secretion.
3. **Interneurons** connect sensory and motor neurons and also carry stimuli in the brain and spinal cord.

Neurons propagate nerve impulses. Such an impulse is an electrochemical phenomenon. In an inactive neuron, a neuron at rest, the cytoplasm is negatively charged with respect to the outside of the cell. When this difference in electrical charge disappears, a nerve impulse is generated. This happens when a stimulus contacts the tip of a dendrite, increasing the permeability of the cell membrane to sodium ions. When this occurs, the ions rush back into the cytoplasm, thereby removing the difference in electrical charges and creating an electrochemical pulse—the nerve impulse.

The nerve impulse travels through the dendrite, through the cell body, and down the axon. At the end of the axon is a fluid-filled space called a *synapse.* Synapses may occur between the axon of one neuron and the dendrite of another or between neurons and muscle fibers, forming a *neuromuscular junction.* When the impulse reaches the end of the axon, it causes the release of chemicals called *neurotransmitters.* Molecules of these substances accumulate at the synapse, increasing the membrane permeability of the next dendrite. This causes a new impulse to be generated in that neuron. Thus the impulse is transmitted from neuron to neuron. As with muscle fibers, impulse transmission either occurs or does not occur. Degree is determined not by the intensity of a particular impulse in a particular neuron, but by the number of neurons that are stimulated.

In human beings, nervous coordination is carried out by two nervous systems, the *central nervous system* (CNS) and the *peripheral nervous system* (PNS). The nerves of the CNS lie within the brain and spinal cord, while those of the PNS extend outside of these structures to and from the CNS.

The *spinal cord* extends from the base of the *brain* to the bottom of the backbone. The spinal cord detects certain stimuli and responds to them directly (without reference to the brain) in what is called a *reflex arc.* The spinal cord also serves as a kind of trunk line, distributing nerves from the brain to the rest of the body.

The brain is the most complex of all human organs. It is a complex mass of nervous tissue and is the site of consciousness, sensation, memory, and intelligence. In general, the superficial portion of the brain, the *cerebral cortex,* controls the higher intellectual functions, while the structures of the inner portion control the body's many autonomic activities.

The peripheral nervous system consists of the *sensory somatic system*, which carries impulses from the external environment and the senses, and the *autonomic nervous system*, which involuntarily prepares the body to meet various threats by increasing the heartbeat, constricting arteries, and dilating pupils.

Besides the brain, spinal cord, and nerves, the senses are the other key organs in the nervous system.

- **The eye.** This organ focuses light on a layer of specialized nerve cells distributed across the *retina*, a structure on the back of the eyeball. These cells generate nerve impulses when stimulated by light—some (called *rods*) detect only light and dark, while others (called *cones)* detect different colors. The impulses are carried from the retina along the *optic nerve* to the brain.
- **The ear.** The ear transforms sound waves into nerve impulses by funneling these waves to the *eardrum*, which vibrates against three tiny bones, the *malleus*, *incus*, and *stapes*. These bones, in turn, transmit the vibrations to the *cochlea*. Fluid within this structure vibrates in response to vibrations transmitted from the three bones. Within the cochlea, these vibrations are picked up by hair cells that excite nerve impulses, which are carried along the *auditory nerve* to the brain.
- **Taste and smell.** These related senses operate through *chemoreceptors*, specialized cells that are stimulated variously by the molecules and ions of substances in the environment. When stimulated, nerve impulses are generated and transmitted via the *olfactory nerve* to the brain.
- **Touch and other senses.** Receptors for touch and pain are called *pacinian corpuscles* and are located in the skin, muscles, and tendons. Changes in pressure cause the generation of nerve impulses. The sense of balance is largely centered in the inner ear, where changes in the orientation of fluid in the *semicircular canals* stimulates hair cells to fire certain nerves. Internal or visceral senses (such as fullness of bladder or bowel) are created by *stretch receptors* in the muscles and by *carbon dioxide receptors* in the arteries.

Endocrine System

Homeostasis requires a nervous system to achieve and maintain the body's nervous coordination, and a chemical system to achieve and maintain *chemical coordination*. Animal bodies, including the human body, use a system of *endocrine glands* to secrete *hormones*, proteins and lipids that bring about changes in the body and help coordinate body systems. The endocrine glands secrete their various hormones directly into the bloodstream, which carries the hormones to *target organs*—the organs the particular hormone is supposed to stimulate. In contrast to such ducted body glands as the salivary glands (which secrete saliva for digestion) and the sweat glands, the endocrine glands have no ducts and are sometimes referred to as the *ductless glands.* (The ducted glands are sometimes called *exocrine glands.*)

Pituitary Gland

In human beings, the *pituitary gland* is located at the base of the brain. It secretes important hormones that play key roles in regulating growth and maturation. One of these hormones, *human growth hormone* (HGH), accelerates protein synthesis to promote body growth. Oversecretion of the hormone produces gigantism, while undersecretion produces dwarfism. *Lactogenic hormone,* also called *prolactin,* promotes breast development in females and stimulates the secretion of milk.

The endocrine glands often work in an interrelated fashion. The pituitary produces *thyroid-stimulating hormone* (TSH), which controls secretions from the thyroid gland. *Adrenocorticotrophic hormone* (ACTH) controls secretions from the adrenal gland.

The pituitary also secretes:

- **Follicle-stimulating hormone** (FSH). In females, this stimulates development of a follicle, which contains an egg cell. In males, it stimulates sperm production.
- **Luteinizing hormone** (LH). In females, LH completes the maturation of the follicle and also stimulates formation of a structure called the *corpus luteum,* which secretes more female hormones.
- **Interstitial cell-stimulating hormone** (ICSH). The male counterpart of LH, it stimulates the secretion of male hormones in the testes.
- **Melanocyte-stimulating hormone** (MSH). Stimulates production of *melanin,* the skin pigment.
- **Antidiuretic hormone** (ADH or vasopressin). Produced in a part of the brain called the *hypothalamus,* this hormone is stored and released by the pituitary. It stimulates water reabsorption by the kidneys.
- **Oxytocin:** Another hormone produced by the hypothalamus and stored and released by the pituitary, oxytocin stimulates contractions of the uterus during the birth process.

Because its hormones regulate so many functions, including the work of other glands, the pituitary is sometimes referred to as the "master gland."

Thyroid Gland

Located (in human beings) against the pharynx, at the base of the neck, the *thyroid gland* produces *thyroxine* and *calcitonin.* Thyroxine regulates the body's metabolic rate. Calcitonin regulates the level of calcium in the blood.

Parathyroid Glands

These structures are found on the posterior (rear) surfaces of the thyroid gland and produce *parathyroid hormone,* or *parathormone.* This substance regulates calcium metabolism.

Adrenal Glands

Located on top of the kidneys, the *adrenal glands* consist of an outer *cortex* and an inner *medulla.* The cortex secretes several substances collectively called *corticosteroids.* These stimulate and regulate a variety of body functions, including mineral metabolism, glucose metabolism, and protein synthesis.

The medulla portion of the adrenal glands secretes *epinephrine* (also called *adrenaline*) and *norepinephrine* (*noradrenaline*). Epinephrine increases the heart rate and blood pressure, thereby increasing the blood supply to the skeletal muscles. In times of fear or excitement, when the individual is threatened, this "adrenaline rush" helps the body prepare for what is frequently called the "fight-or-flight response." Increased blood supply to the skeletal muscles enables more effective defense in the form of fighting or fleeing. Norepinephrine serves to augment the effects of epinephrine.

Pancreas

The human *pancreas* is located behind the stomach. Cell clusters called the *islets of Langerhans* secrete two important endocrine hormones, *insulin* and *glucagon.* Insulin promotes passage of glucose molecules into the body and regulates their metabolism. Without insulin, glucose is simply removed from the blood and excreted, resulting in diabetes mellitus, a disease characterized by extremely sluggish body metabolism.

Glucagon stimulates the breakdown of glycogen to glucose in the liver. It also releases fat from *adipose tissue* (tissue where fat is stored) so that it can be used for the synthesis of carbohydrates.

Ovaries and Testes

In females, the *ovaries* secrete *estrogens,* which promote development of secondary female characteristics. In males, the *testes* secrete *androgens,* which promote secondary male characteristics.

Pineal Gland

Located in the midbrain, this small gland is something of a mystery, though it seems to play a role in regulating mating behavior and in regulating the body's day-night cycles.

Thymus Gland

Found in the neck tissues, the *thymus* secretes *thymosins*, which act on the immune system to stimulate T-lymphocytes.

Other Endocrine Secretions

Various body tissues secrete *prostaglandins*, which have diverse effects. Various prostaglandins stimulate smooth-muscle contraction, lower or raise blood pressure, decrease and increase the clotting ability of blood, enhance ion transport across some membranes, stimulate inflammation, and inhibit the breakdown of fat in adipose tissue.

Kidney cells produce *erythropoietin*, which plays a role in the production of red blood cells, and digestive glands secrete *gastrin* and *secretin*, which aid in regulating the digestive process.

Reproductive System

The reproductive system produces *gametes—eggs* in the female and *sperm* in the male. It then prepares the gametes for fertilization. In the male, the reproductive system also functions to deliver the sperm to the female. Her reproductive organs must nourish and protect the fertilized egg as it develops into an embryo, fetus, and baby.

The Male Reproductive System

The focal organs of the male reproductive system are the testes, located in the scrotum, which, lying outside the body cavity, keeps the temperature of the testes somewhat below that of normal body temperature. This is important to the development of viable sperm within the coiled *seminiferous tubules* of the testes. The sperm cells, each bearing a haploid number of chromosomes, mature in the *epididymis*, located along the surface of the testes. Sperm is delivered from the epididymis to the female through the *urethra* of the *penis*. When sexually aroused, the penis becomes erect as blood fills spongelike tissue within it, the sperm cells are mixed with secretions from the *prostate gland*, the *seminal vesicles*, and *Cowper's glands*, becoming semen, which is ejaculated into the female during intercourse.

The Female Reproductive System

The female reproductive system consists of the *ovaries*, two egg-bearing organs within the pelvic cavity, two *Fallopian tubes* (oviducts), the *uterus*, and the *vagina*. In human beings, the production of eggs actually begins before birth. Some two million *oogonia*, primitive egg cells, accumulate in the ovaries and develop into *oocytes* after puberty. Each month, an oocyte develops into an egg cell within an ovarian structure called the *Graafian follicle*. After about fourteen days of development within this structure, the follicle releases the egg cell (the process is called *ovulation*), and the egg cell moves through the Fallopian tube toward the uterus. While this occurs, the follicle, acted on by luteinizing hormone (LH) secreted by the pituitary gland, is transformed into a structure called the *corpus luteum*, which secretes progesterone. Together with estrogen, this hormone prepares the *endometrial tissue* of the uterus for implantation of a fertilized egg.

The egg remains viable in the Fallopian tube for twenty-four to seventy-two hours. If fertilized by a sperm cell during this time, it will become a *zygote*, bearing the full human diploid complement of forty-six chromosomes. The zygote undergoes mitosis, becoming a ball of cells called a *morula*, which moves along the Fallopian tube. The morula develops into a *blastocyst*, a hollow ball of cells, at about the time that it enters the uterus. The blastocyst becomes implanted in the endometrial wall of the uterus, and the blastocyst's outer layer of cells (the *trophoblast*) begins to form the *placenta*, the structure that will nourish and support the developing embryo and fetus. The trophoblast also develops the *amnion*, *chorion*, and *yolk sac membrane*.

In the meantime, the inner mass of cells of the blastocyst differentiates into the three germ layers common to chordates: the ectoderm, mesoderm, and endoderm. By the fourth week after fertilization, the blastocyst is developed into an *embryo*. Development proceeds, and by eight weeks, the embryo is a *fetus*, with many of the obvious anatomical features of a human being.

In human beings, the embryo and fetus develop over the course of nine months, at which point various hormones operate to stimulate the beginning of the birth process.

In the event that an egg cell in the Fallopian tube is not fertilized, the egg moves toward the uterus, and the corpus luteum begins to break down. Within two weeks, hormone levels in the body decline so that they do not inhibit uterine contractions. Contractions occur, and the degenerated endometrium sloughs off and is expelled by the uterus in a process called *menstruation*. The cycle of follicle development then begins anew, but in the opposite ovary.

CHAPTER 16

GENETICS

THE DEFINING PROPERTY OF A SPECIES IS THAT MEMBERS SHARE A SET OF INHERITED characteristics. Yet, within a species, there may be wide variation. Individuals inherit characteristics that are common to the species, as well as characteristics in varying degrees unique to the individual. The inherited characteristics—called *traits*—of an organism are transmitted to that organism by genes. How the genes do this and how the traits are inherited is the province of the branch of biology called *genetics.*

Classical Genetics

The science of genetics advanced greatly beginning in the 1950s and 1960s with the discovery of how the molecule DNA operates as the central molecular mechanism of genetic inheritance. However, the study of genetics did not *start* on this molecular level. So-called *classical genetics* began in the 1860s and 1870s with the work of Gregor Mendel (1822-1884), a humble Austrian monk. Mendel experimented with pea plants, focusing on patterns of inheritance of seven obviously varied and visible traits: plant height, pod shape, pod color, seed shape, seed color, flower color, and the location of the flower on the plant. Mendel chose well in focusing on pea plants, because they are self-pollinating, developing individuals that are *homozygous* for particular characteristics.

The different forms of a gene are called *alleles.* A pea plant (for example) may have an allele for tall plants and one for short plants. A tall *homozygous* pea plant will have two tall alleles. Its line of inheritance for this characteristic is said to be pure.

Mendel took various *pure-line* pea plants and artificially cross-pollinated them with other pure-line plants, then noted the resulting offspring. After years of patient experimentation, Mendel formulated three cornerstone laws of inheritance:

1. **Mendel's Law of Dominance.** When an organism has two *different* alleles for a trait, one allele dominates (and is called a *dominant trait*). Nondominant alleles are called *recessive traits.*
2. **Mendel's Law of Segregation.** During gamete formation by a diploid organism, the pair of alleles for a given trait *segregates* (separates)—as in meiosis. If hybrid (*heterozygous*) individuals are crossed, the recessive trait segregates out at a ratio of 3 to 1; that is, for every three offspring exhibiting the dominant trait, one will exhibit the recessive trait.
3. **Mendel's Law of Independent Assortment.** During meiosis, the members of a gene pair separate from one another independently of the members of other gene pairs. That is, each trait is inherited independently of others.

The Gene Concept

Mendel spoke of "factors," not genes. He neither studied nor speculated on the actual mechanisms of inheritance; he only observed—and brilliantly so—the outward workings of inheritance. In 1911, the biologist Thomas Hunt Morgan (1866-1945) began to work with fruit flies to study the mechanisms of inheritance more closely. He concluded that the *chromosome* is the means by which hereditary traits are passed from one generation to the next.

Recall that sex cells replicate through meiosis. During this process, the threadlike cellular structures called chromosomes duplicate and pass into two cells. The chromosomes of the two cells then separate and pass into four daughter cells. The parent cell has two sets of chromosomes (so is diploid), while the daughter cells each have a single set of chromosomes (and so are haploid). These daughter cells are the gametes, the sex cells that will fertilize or be fertilized by the gametes of another organism (of the same species), creating a diploid zygote, which will develop into an embryo and, ultimately, into a new organism. The new organism combines chromosome-borne traits of both of its parents.

Through his work with fruit flies, Morgan further concluded that chromosomes are composed of smaller, discrete units he called *genes.* It is the genes that actually carry specific traits. The genes move with the chromosomes during the process of cell division. (Morgan also proposed that, if the genes change—mutate—the traits they control change.)

Punnett Squares

In most introductory biology surveys, you will likely be asked to work some genetic problems based on the concepts of Mendel and Morgan. The basic tool for working such problems is the Punnett square, which is used to predict the *genotypic* and *phenotypic* ratios that result from the combination of various gametes.

NOTE: ***Genotype*** **is all the genetic traits of an organism, as revealed by breeding experiments. The** ***phenotype*** **is the visible traits of the organism—the visible expression of genotype—which can be seen in the organism without any breeding experiments.**

Here is a simple version of one of Morgan's fruit fly experiments. Some fruit flies exhibit long wings. Others exhibit short, vestigial wings. The trait for long wings is dominant—and is, therefore, expressed in "shorthand" form with a capital letter: L. The trait for vestigial wings is recessive, and is expressed with a lowercase letter: l. A fruit fly is *heterozygous* for wing length if it bears genes for both long (L) and vestigial (l) wings. What happens if two such fruit flies mate? You can sketch a Punnett square to find out:

Male × Female

Ll Ll

	L	l
L	LL	Ll
l	Ll	ll

Based on the Punnett square, you may conclude that the next generation of fruit flies will consist of:

1 homozygous dominant long-winged fly (LL)
2 heterozygous dominant long-winged flies (Ll)
1 homozygous recessive short-winged fly (ll)

Note that these results list ratios of ***genotype.*** **The** ***phenotypes*** **(visible characteristics) represented here are three long-winged flies to one short-winged fly. You cannot tell by looking at the long-winged flies whether they are homozygous for the trait or heterozygous for it.**

A Loophole in Mendel's Law of Dominance

There are numerous instances in which Mendel's Law of Dominance does not hold true. This is the case with *codominance* (sometimes called *blending*), when neither of two traits is dominant, but both are inherited. In some flowers, the hybrid of, say, a white-flowered plant and a red-flowered plant will be pink, a blend of the white and red traits.

Crossovers and Mutations

Theoretically, genes are linked on chromosomes, so they are inherited as a group on a particular chromosome. In actuality, linked groups may be broken during meiosis, and chromosomes may exchange homologous parts in a phenomenon called *crossing over.*

It is also possible that genes can be changed through *mutation.* Mutations are usually random—though they may also be caused by certain chemicals and by radiation. Mutations are usually recessive, and they are usually (but not always) harmful.

Deletion is a type of mutation caused by the loss of a piece of chromosome. *Duplication* results when a piece of a chromosome adheres to another chromosome, giving that chromosome too many genes. In the case of *inversion*, genes become rearranged on a chromosome, thereby changing the genetic sequence on the chromosome and preventing chromosomes from matching up properly during meiosis. Finally, individual genes can change due to *point mutations.*

Another type of mutation, polypoidy, results when cells develop extra sets of chromosomes. This condition is sometimes purposely induced with chemicals to create polypoid fruits and flowers, which are very large.

Sex-Linked Traits

In human beings, one pair of chromosomes determines the sex of the offspring. In females, this pair is designated XX, while in males it is designated XY. The Y chromosome is much shorter than the X chromosome. This means that when a gene occurs on one X chromosome in a female, a corresponding gene also probably occurs on the other X chromosome of the pair. However, in the male, such a corresponding gene probably will not be present on the short Y chromosome. For example, the trait of red-green color blindness is recessive and is carried on the X chromosome, as is the dominant gene for normal vision. If one of the female's X chromosomes has the gene for color blindness, the chances are that she will have the dominant gene for normal color vision on her other X chromosome. Thus very few females are color blind. If, however, the color blindness gene is present on a male's chromosome, the dominant gene for normal color vision will not be present on the Y chromosome, because it is so short. Males, therefore, have a much greater chance of inheriting the color blindness phenotype.

Traits such as color blindness and hemophilia are said to be sex-linked, because the genes for them occur on the X chromosome.

Action of Multiple Alleles

According to classical genetic theory, *alleles* are two or more genes that occupy the same positions on homologous chromosomes. Alleles are separated from each other during meiosis, then come together again at fertilization, when homologous alleles are paired—one from the sperm cell and one from the egg cell. Yet the inheritance of some traits cannot be explained by the action of a single pair of alleles. Sometimes multiple alleles are involved, as in the inheritance of blood type in human beings:

Blood type A may be produced by genotype I^AI^A or I^Ai
Blood type B may be produced by genotype I^BI^B or I^Bi
Blood type AB may be produced only by genotype $I^A I^B$
Blood type O may be produced only by genotype ii

Molecular Genetics

The field of genetics was revolutionized beginning in 1953, when biochemists James D. Watson and Francis H. C. Crick described the structure of the deoxyribonucleic acid (DNA) molecule as two nucleotide chains that twist around one another to form a double helix, rather like a spirally twisted ladder. The curved sides of this ladder consist of alternate phosphate-sugar groups, while the rungs of the ladder are protein bases joined by weak hydrogen bonds. DNA can replicate itself, which it does prior to mitosis or meiosis. The double-stranded helix unwinds, and each "unzipped" strand picks up complementary nucleotides, which, incorporated into the unzipped strands, create two new DNA molecules that are identical to each other and to the original molecule.

The DNA molecule carries chemically coded instructions for controlling all cell functions. During meiosis or mitosis, another nucleic acid molecule, RNA, carries this code from the nucleus of the cell out to the cytoplasm, where proteins are synthesized. This synthesis proceeds according to the code embodied within the DNA.

The type of RNA that carries the DNA code to the cytoplasm is called *messenger RNA* (mRNA). Another type of RNA, *ribosomal RNA* (rRNA), is used in the creation of ribosomes, particles of rRNA and protein that are the sites of protein synthesis. Yet another type of RNA, *transfer RNA* (tRNA), exists in the cell cytoplasm and carries amino acids to the ribosomes for protein synthesis.

Depending on the depth, breadth, and focus of the biology course you are taking, you may be exposed to a general overview of the highly complex process by which the DNA's genetic code is transferred, transcribed, and translated into protein synthesis in the ribosomes, or you may explore the process in detail. You may also proceed from a survey of genetics on the molecular level to a consideration of biotechnology—genetic engineering, which can manipulate DNA (as so-called *recombinant DNA*) to form proteins not normally produced in a cell. Biotechnology can be used to create new drugs, vaccines, and other products. It takes genetics from an essentially descriptive and theoretical discipline to an applied science (and one that is often controversial).

CHAPTER 17

EVOLUTIONARY THEORY

IT DOES NOT REQUIRE MORE THAN AN AVERAGE DEGREE OF CURIOSITY TO LOOK AROUND oneself and ask what is responsible for the great variety of living things. Virtually all religious traditions incorporate creation stories. The biblical Old Testament, for example, begins with a brief description of how God created all matter and all beings from nothingness during the space of six days. Yet fossils and other evidence suggest that creation was not such a quick event and that life, in all its variety, has evolved (and continues to evolve) over a long time.

Two Theories of Evolution

In the late eighteenth century, the French biologist Jean Baptiste Lamarck (1744-1829) proposed that the great variation among living things was the result of changes stimulated by the environment. If an organism's environment made a certain demand, systems and body structures would develop in response to that demand. For example, a giraffe's neck might stretch in order to reach the topmost leaves on the trees it uses for food. Moreover, this *acquired characteristic*, a stretched neck, would be passed on to the giraffe's offspring. Over many generations, giraffe necks would, therefore, become quite long. If, on the other hand, it somehow became environmentally unnecessary for the giraffe to have a long neck, not only would these animals stop stretching, but this body feature would eventually disappear through disuse. Like the acquisition of the long neck, that disappearance would pass from generation to generation.

Lamarck developed his theory well beyond these simple points, and he drew a great many followers; however, it is fairly easy to demonstrate that acquired characteristics are not passed from one generation to the next. Another biologist, August Weismann (1834-1914), cut off the tails of mice, then mated them. He repeated this for twenty-five generations. The mice of the twenty-fifth generation grew tails that were just as long as those of the first generation. There had been no measurable acquisition of the characteristic of taillessness.

In 1831, twenty-two-year-old Charles Darwin (1809-1882), an English naturalist, sailed on H.M.S. *Beagle*, collecting, over a five-year period, many animal and plant specimens and making many acute and varied observations. For years afterward, he studied and collated what he had observed, writing, in 1859, a book entitled *On the Origin of Species by Means of Natural Selection, or the Preservation of Favored Races in the Struggle for Life.*

The title and subtitle of the volume summarize Darwin's controlling idea: Only organisms best adapted to their environmental circumstances survive the struggle for life and produce offspring; those that are not well adapted do not survive; eventually, their line ends, and the species becomes extinct. The surviving species, in all their variety, are the plants and animals we know.

Darwin's theory is founded on three broad premises:

1. Organisms produce more gametes or offspring than can possibly survive.
2. Organisms fail to survive because they must compete for scarce food and territory. They must struggle to survive. Those least suited to the struggle die. Those best suited to it live—and reproduce.
3. Survival is a matter of competitive edge. Variations—in structure, in agility, in keenness of senses, in food requirements, and so on—exist between species and between members of the same species. On average, those species and individual organisms that possess variations that better suit them to compete will survive and reproduce. On average, those species and organisms that do not possess these variations will die and fail to reproduce. Over time, the most useful variations will tend to be perpetuated and developed because more individuals possessing these variations will survive to reproduce than individuals that do not possess these variations.

Beyond Darwin

Darwin's theory of natural selection—that the pressure of the environment effectively "selects" the traits that are perpetuated in a species—was and is momentous in its implications and in its effect on biology. Nevertheless, Darwin failed to explain the genetic basis for variations. Indeed, he not only assumed that all variations that are of value to survival are passed from parent to offspring, but that (as Lamarck had held) acquired characteristics are inherited.

Subsequent biologists have worked to explain the mechanisms of variation.

Mutation

Mutations (see Chapter 16) are random changes in the *gene pool* (the sum of genes of a particular population) due to a change in the DNA or in one or more chromosomes. Many mutations are harmful, leading to the death of the organism and, therefore, a failure of the mutation to be repeated; however, some mutations are beneficial—an aid to survival. In this case, over time, mutated individuals survive, reproduce, and the mutation enters and alters the gene pool.

Gene Flow and Genetic Drift

Evolution is in part a function of geographic distribution. Darwin, for example, closely studied the species of the Galapagos Islands, many of which were exclusive to the region. Geographic isolation had allowed certain unique species to evolve. When migrating individuals enter a geographic region and interbreed with the existing population, they introduce new genes into the gene pool, which may include beneficial variations. This is called *gene flow.*

Genetic drift is the complement of gene flow and occurs when a group of individuals leaves a larger population and starts a new group in some geographically isolated region. The new group will evolve differently from the larger, original group.

Natural Selection Revisited

Biologists still believe that natural selection, the process Darwin described, is the most important influence on evolution. However, they have developed the concept beyond Darwin's original formulation and have rejected the notion of the inheritance of acquired characteristics. It is sufficient that extremely slight variation exists in a population. Environmental conditions will favor certain variations, no matter how slight, over others. A few more of the favored individuals will survive to reproduce. Thus, over time, the numbers of nonfavored individuals will dwindle and disappear, and the species will be defined by the favored variation. Over a very long time—over many generations—the variations may become more pronounced.

A Matter of Time

Darwin argued that evolutionary change takes place very gradually. This is often called *gradualism.* Some biologists argue that some species have very long periods of stability interrupted by relatively brief periods of rapid change in response to some environmental or geological change. This is called *punctuated equilibrium.*

CHAPTER 18

ECOLOGY

THE WORD *ECOLOGY* WAS COINED IN 1869 BY A GERMAN ZOOLOGIST, BUT INTENSE interest in ecology—the relationship between living things and their environment—has been largely a more recent phenomenon, spurred in great measure by a worldwide conservation movement. As human populations have grown and civilization has become increasingly industrialized, many people have come to the realization that the demands of that growing population and the by-products of industrial civilization—mainly pollution of air, water, and soil—pose a danger to various animal and plant populations. The desire to preserve such populations has led to a need to understand how organism populations relate to everything that surrounds them. This, the *environment*, includes *biotic* as well as *abiotic* factors. With regard to a particular plant or animal population, biotic factors are other populations of organisms that share their environment. Abiotic factors are the physical and chemical conditions that prevail in the population's environment.

Thinking in Terms of Ecosystems

The study of ecology is the study of systems more than individual organisms or particular biotic or abiotic factors. Typically, an ecologist will define a particular *population* to study (the members of a given species in a given location) or will define a particular *community* (all of the plant and animal populations within a given environment). The interaction of the community and the nonliving environment is an *ecosystem*. Any place in which the interaction of the community and the nonliving environment can be usefully studied can be defined as an ecosystem. This might be a vast tract of ocean, a great forest, or small pond, a vacant city lot, or even a mound of trash or a few square inches of dirt.

A typical ecosystem consists of:

- **The physical environment**—air, water, soil.
- **The *producers.*** In many ecosystems, these are the green plants. They are defined as producers because they are autotrophs—they produce their own food, using abiotic factors.
- **The *primary consumers.*** These are herbivore animals—organisms that depend directly on the producers for food.
- **The *secondary consumers.*** These are carnivores, which feed on the herbivores, who feed on the producers.
- **The *tertiary consumers.*** These are animals that feed on smaller carnivores as well as herbivores.
- **The *scavengers*—**animals that feed on dead organic matter, including the bodies of dead animals.
- **The *decomposers*—**bacteria and fungi that break down dead organic matter, thereby releasing from it organic compounds that are recycled into the soil. It is this recycled material that the producers use to synthesize their food.

An ecosystem involves much interaction and competition, but it is ultimately cyclical, as can be seen by the relationship between the producers and the decomposers.

Niches and Competition

Some members of a community live in mutually beneficial relationships. Some live in parasitical relationships. Some live in more or less indifferent relationships of peaceful coexistence. Some live in direct competition with one another. An *ecological niche* is a role or way of life within a particular community. Two species may live close to one another, and they may even eat the same kinds of food, but if they feed in significantly different ways or in even slightly different locations, they may be said to occupy different niches and, therefore, do not compete with each other. If, however, two species occupy the same exact place and eat the same food, they are in competition. The species better adapted to obtaining this food will survive.

Another Way of Looking at the Ecosystem

Ecosystems can be regarded from the point of view of communities, niches, and competition, or from the point of view of energy conversion. Remember, all living things carry out metabolism: they use energy to drive the processes of life. The ultimate source of energy in any ecosystem is the sun. The producers, the green plants, transform a fraction of the sun's energy into food (in effect, stored energy), which the consumers use.

Food Chains

How can one study the flow of energy through an ecosystem? By examining how stored energy—food—is consumed. A green plant feeds an herbivore animal, which, in turn, falls prey to a small carnivore, which, ultimately, is consumed by a larger carnivore. That animal leaves some parts of its prey unconsumed. This is eaten by scavenger animals. Eventually, even the tertiary consumer at the end of this *food chain* dies. Its remains are broken down by decomposers, who return the organic compounds and minerals to the soil, which nourishes the green plants (the producers).

The story of the food chain is a story of energy, with the greatest amount concentrated in the producers and decreased at each step, only to be reconcentrated in the end by the decomposers.

Of course, living systems are rarely quite this simple. The food chain is a straight-line, highly schematic, and quite flat model. A three-dimensional model would more accurately reflect reality. Ecologists usually think in terms of a *food web* rather than a food chain, since few animals eat only *one* other kind of animal or are, in turn, the prey of only *one* other predator. Food and feeding relationships resemble less a straight line than they do the pattern of a web. Indeed, it may be even more useful to look beyond particular species and beyond *exactly* who is eating whom. Instead of a web, one can picture energy flow in an ecosystem as a pyramid. In such a picture, all of the producers are lumped at the base, with all herbivores above them as the next slab of the pyramid, then the first-level carnivores above them, and the second-level carnivores at the top. Consider the shape of this *food pyramid.* The autotrophic producers really do *support* all of the levels of organisms above them. The narrowing of the pyramid reflects the progressive decrease of available energy at each step.

Abiotic Cycles

Yet another way to study the ecosystem is from the point of view of the abiotic factors that enter into it. These are sometimes referred to as *biogeochemical cycles.* They include the water cycle, the carbon dioxide/oxygen cycle, and the nitrogen cycle. Each of these can be usefully traced through the ecosystem. For example, with the water cycle, one could begin with rain being absorbed by the soil, from which plant roots pull it into plants, where it is used in the conversion of sunlight into oxygen (used by all consumers in respiration) and sugars, which are consumed by the primary consumers (and, through them, ultimately by the other consumer levels). All of the consumers exhale carbon dioxide, which the green plants require, as well as water vapor, which is ultimately condensed in the atmosphere and returned to the earth (and the ecosystem) as rain.

The Limits

Once all of the cycles of a particular ecosystem are understood, it becomes possible to study the factors that *limit* the growth and survival of species in that ecosystem. A given ecosystem will support only so many organisms of this or that type. The amount of available sunlight, water and rainfall, temperature, and so on determines in varying ways the survivability of particular species in particular ecosystems. Understanding *limiting factors* and their effects on particular species under given circumstances is important in conservation efforts. It also provides insight into certain aspects of evolution and species adaptation.

Succession

One problem with viewing ecosystems in terms of cyclical relationships is that it is easy to lose sight of change. The fact is that, over time, physical conditions prevailing in an ecosystem do change and so do the organisms in the ecosystem. The orderly changes within the biotic community of an ecosystem are called *ecological succession.*

Primary succession takes place when living things first come to occupy a formerly barren place. How does this happen? Let us take a rocky region. Through time, rocks erode and weather. When the rock is sufficiently pulverized, lichens may find a foothold and grow. These organisms may be the first living things in this region. The lichens secrete chemicals that hasten the breakdown of the rock into soil, and they also contribute organic matter to the mix. At some point, the soil may become hospitable to mosses, which add more organic matter as well as water-retaining ability. Grasses follow, then bushes, taller shrubs, and eventually trees. Each population creates conditions that allow the next population to enter the ecosystem and survive.

Primary succession is said to yield to a *climax community*—a stable ecosystem—when one or two species of large trees predominate. *Secondary succession* may occur when one species of predominant tree is replaced by another.

Biomes

A climax community that extends over a vast geographical region characterized by one type of climate is called a *biome.* The world may be divided into seven major biomes:

1. **Tundra:** treeless plains surrounding the Arctic Ocean

2. **Taiga:** subarctic coniferous forest region
3. **Deciduous forest:** temperate-region forests
4. **Desert:** regions of limited rainfall (<6.5 cm per year) or areas where rain occurs unevenly or is subject to a high rate of evaporation
5. **Grasslands:** regions of low and irregular rainfall
6. **Tropical rain forest:** regions of high temperature and constant rainfall
7. **The sea:** some three-fourths of the earth's surface

Each of these biomes has characteristic plant and animal populations determined by characteristic limiting factors.

The Human Element

Many introductory biology courses include units on the effect of human activity on the *biosphere*—the realm of living things—and on conservation. Some introductory courses are focused on this crucial topic and use it as an entryway into the more basic and broader principles of biology.

PART THREE

MIDTERMS & FINALS

CHAPTER 19

BUFFALO STATE COLLEGE

BIOLOGY 100: PRINCIPLES OF BIOLOGY

Javier Peñalosa, Associate Professor

GENERALLY, THERE ARE THREE EXAMS IN THE COURSE, ALL EQUALLY WEIGHTED. TWO are reproduced here. Usually, the course grade is based entirely on the exams. My exams are mostly "short-answer" exams.

The main course objective is to give students an appreciation of the principles of biology as they apply at the cellular and molecular level. All of the students are nonmajors. I want students to make connections and understand processes. This is more important than memorizing all the steps in a process—for example, knowing the Krebs cycle cold. Since I have to assume that this is my students' last (as well as their first) biology course, I don't worry too much about what they will "need to know" for further work in biology. Thus I hope the exams reward understanding rather than memorization.

Success in the course requires attending class, copying your own notes, and preparing study notes. Don't use a highlighter. Instead, use the text as it was designed to be used: answer the review questions, study and understand the figures and tables, and be alert to the section headings and summary statements.

Many students claim to be "visual learners," but they tend to ignore illustrations.

Study in small groups regularly—perhaps once a week at the same time always. Ask questions in class. Send me e-mail with questions. Come to the review session prepared. Don't expect others to ask questions for you.

EXAM 1

1. What are genes? (Two definitions)

Answer:

1. Genes are bits of information that specify traits.
2. Genes are segments of DNA that contain the information for synthesis of specific proteins.

This is a simple definition recall question, but it requires students to understand the gene concept on two levels.

2. Why do Hoagland and Dodson use the metaphor of the "robot" to describe protein molecules?

Answer: The robot proteins are enzymes. These molecules, like robots, are programmed to perform the same task over and over.

I used the illustrated book Hoagland and Dodson, *The Way Life Works*, for this session. Students carefully study the cartoonlike illustrations, which are very information rich. The robot metaphor stresses the automatic functionality of the enzyme as determined by the shape of its active site.

3. What makes one kind of amino acid different from another kind?

Answer: Different amino acids have different side-chains ("R-groups").

A chemistry question. I tried to keep these to a minimum, but I believe that this was essential. The book and I "taught" this idea by displaying illustrations of amino structure, showing that they all have the same functional groups, but that they differ in their side-chains. An easy question for the moderately attentive student.

4. The molecules CO_2 (carbon dioxide) and CH_4 (methane) are about the size of H_2O, but (unlike H_2O) are gases at room temperature. What accounts for the ability of water to remain a liquid at room temperature?

Answer: Because of the polarity of the H_2O molecule, weak hydrogen bonds form between H_2O molecules, helping to keep water together. These bonds do not form between CO_2 and CH_4 molecules.

Students found this difficult. The question is heavily theory laden: molecular weight, solid/liquid/gas states, charge and polarity, molecular structure, chemical bonds, and (most importantly) the idea that the macroscopic properties of a familiar substance are explicable on the atomic level. Not all of the students are convinced that atoms are real. After all, they can't see them!

5. Name the two main uses of sugar molecules in organisms.

Answer: Sugar is used as a source of energy to make ATP during cellular respiration and as a "building block" in the synthesis of more complex compounds, e.g., amino acids and nucleotides.

The idea that a molecule "contains" energy is a tricky one that I have difficulty teaching. Students were encouraged to think about what happens to our food after we eat it. How does a baby gain weight? If you kept track of the weight of food a baby ate, and the baby's increase in weight, would the quantities be the same?

6. Give two examples of chemical compounds that are a waste product of one chemical reaction, but required for another reaction.

Answer: CO_2 and H_2O are required for photosynthesis, but are waste products of cellular respiration. O_2 is required for cellular respiration but is a waste product of photosynthesis.

This isn't a very good question. I wanted to stress the idea of the interconnectedness of biochemical cycles, how the output of one cycle becomes the input of another. I need a more straightforward way of getting at this.

7. Organisms use energy for growth, reproduction, storage, and "maintenance." Explain why "maintenance" requires energy.

Answer: Maintenance depends on the continual breakdown and resynthesis of molecules, membranes, cells, etc. These processes require energy.

Hoagland and Dodson use the image of a cathedral that is being rebuilt—part by part—as it is being torn down. The students have to keep this metaphor in mind, remember the "robots" (enzymes) doing the work, and then recall that this work requires expenditure of ATP. Lots of ideas to keep in mind and connect.

8. Explain how animals and plants get their amino acids.

Answer: Animals get their amino acids by digesting (breaking down) protein molecules in their food. Plants make their amino acids "from scratch," using photosynthetically produced sugar and nitrogen absorbed from the soil.

To answer this question, a student must recall that amino acids are components of proteins, that they contain nitrogen, that animals eat but plants don't. Once these elementary concepts are recalled, the answer results automatically. I mean that this is the only answer compatible with what the student already knows.

9. Male-male combats during mating season rarely result in serious injury. Explain why not.

Answer: Serious injury would reduce a male's chances of reproducing successfully. Natural selection removes the "super aggressiveness" trait from the population, because males who have it leave behind fewer offspring than those who do not.

Not much inference required here. The book and I said this. Now we ask the student to repeat it.

10. Describe how mitochondria could have evolved from parasites of bacterial cells.

Answer: The ancestor of the modern mitochondrion was a parasitic bacterium, which invaded a larger bacterial cell. Over time a parasitic relationship evolved into a cooperative one as each organism came to depend on the other. The mitochondrion donates ATP to its host, and the host provides compounds for respiration by the mitochondrion.

This is tougher. For one, the student must remember what mitochondria are, what they do, and why it might be advantageous to a cell to have an ATP-producing internal partner.

11. What's wrong with this statement? "Animals respire but plants photosynthesize."

Answer: It's misleading, because plants respire as well. They respire for the same reason animals do: to transfer energy from sugar to ATP.

Students hate this question, identifying it as a trick question. It does require that students replace a prior understanding of "respiration" (= breathing) with the technical definition of cellular respiration (= oxidizing sugar to make ATP), and to recognize that nearly all cells have this need. Students find this idea challenging and resist it. Replacing a prior naive theory with a more sophisticated one requires more mental stamina than many instructors realize.

12. Explain the fact that elephants have wrinkled skin. (Two reasons.)

Answer:

1. A mutation arose that resulted in increased skin growth.
2. The mutation was favored by natural selection because wrinkled skin enhances cooling, especially of large mammals.

This tests understanding of the link between mutation (source of variation in populations) and natural selection (keep good alleles, discard bad ones). Both processes must occur for evolution to occur. Students tend to think that evolution is caused by mutations, and this belief (another prior naive theory) persists, even after I've taught them about natural selection.

13. What is the evidence that oxygen production is not the main function of photosynthesis?

Answer: Some bacteria photosynthesize (produce sugar from CO_2), but make no oxygen.

> Another test of the replacement of a prior notion—in this case, that the purpose of photosynthesis is the production of oxygen.

14. Give an example of "coevolution." Any example from pp. 32-33 [of the text] is okay.

Answer: host-parasite

> This is a simple recall question. I used a loose definition of coevolution. Host-parasite, flower-pollinator, and so on—all would be acceptable.

EXAM 2

Part 1—short answers

1. Hoagland and Dodson say that when chemical bonds are broken, the energy is released as heat. Bearing in mind that in our bodies millions of chemical bonds are being broken every second, why don't we heat up to the point that we catch fire and burn?

Answer: As bonds are broken, much of the energy released is used to form new bonds, not released as heat.

> This tests understanding of the nature and use of bond energy. The question probably needs rephrasing, however, as it implies a contradiction, which really isn't there. Certainly, the text doesn't imply that all bond rupture results in heat alone.

2. If the total amount of entropy (disorder) is always increasing, how can life (which produces order) exist?

Answer: Life uses energy to organize and repair, overcoming entropy.

> This tests understanding of the second law of thermodynamics—or, at least, an understanding of the second law is helpful in answering the question. The question poses an apparent problem with the second law, then says that the problem really isn't one, but asks the student to explain why the problem goes away upon reflection.

3. How is ATP used to "energize" or "prime" a chemical compound to make it more reactive?

Answer: One of the three phosphates of the ATP is attached to the compound.

A simple recall question.

4. What are enzymes and what do they do?

Answer: Enzymes are large protein molecules that catalyze (speed up) chemical reactions in the cell.

Another recall question. Either you know this or you don't.

5. During cellular respiration, energy is transferred from sugar to ATP. Does all of the energy in the sugar end up in the ATP? Explain.

Answer: No. Some of it is lost as heat, in accord with the second law of thermodynamics. The second law says that no process can be 100 percent efficient.

Another test of understanding of the second law. This is a better way of testing that understanding than asking "state the second law of thermodynamics." An answer that says that no reaction can be 100 percent efficient will not get as much credit as one that actually cites the second law.

6. During the Krebs cycle of cellular respiration, high-energy or "hot" electrons (along with protons) are stripped from sugar molecules. How did these electrons get to be "hot"?

Answer: The sugars broken down during respiration were produced by photosynthesis. During photosynthesis, light energy excites electrons, which are ultimately transferred to sugar molecules.

A good answer requires an understanding of the link between photosynthesis and respiration.

7. What is oxygen used for during cellular respiration?

Answer: Oxygen receives the spent electrons from the electron transport system. Hydrogen ions accompany the electrons, so water is produced.

Students remember the "formula" for respiration and memorize the Krebs cycle and electron transport system, but fail to connect the two. I encourage this connection in my lectures and discussion, and I test their ability to make the connection with this question.

8. What is water used for during photosynthesis?
Answer: During photosynthesis water molecules are split to produce oxygen, electrons, and hydrogen ions. The electrons replace the excited electrons transferred from chlorophyll to the electron transport system.

9. What is light used for during photosynthesis?
Answer: To excite electrons in chlorophyll.

The next few questions, 10 through 13, all have the same structure and level of difficulty. They require that students explain respiration and photosynthesis, keeping track of what goes where and where it comes from. Group studying is good preparation for these questions. Students can trace the flow of electrons, compounds, etc., throughout the processes.

10. How are photosynthesis and respiration similar in the way ATP is produced?
Answer: In both processes hydrogen ions become concentrated on one side of a membrane. ATPase enzymes are embedded in the chloroplast and mitochondrial membrane. As the hydrogen ions rush through the channels in the ATPases, ATP is produced.

11. What happens to the ATP produced during photosynthesis?
Answer: It is used in the stroma to pay for the production of sugar.

12. What happens to the ATP produced during cellular respiration?
Answer: It moves out to the cell proper, where it can be used as a source of energy for any cellular process requiring energy.

13. During cellular respiration, CO_2 is produced. Where does the CO_2 come from?
Answer: As the sugar molecules are broken down, the carbon and oxygen in them are released as CO_2.

14. Trivia question: Name the most abundant protein in the world and indicate its function.
Answer: The enzyme ribulose bis-phosphate carboxylase (RuBisCo). It catalyzes the formation of a 6-carbon sugar from ribulose bis-phosphate and CO_2.

This was a giveaway, but the question could be rephrased to elicit a more thoughtful answer, e.g., pointing out that most of the organic matter in the world is plant matter (green), that all leaves contain RuBisCo, etc.

Part 2—essay

15. I have used the concept of "localization" to stress that particular processes (and particular parts of processes) occur in particular parts of organisms. Discuss this concept as it applies either to cellular respiration or to photosynthesis (choose one).

Answers:

Photosynthesis. Photosynthesis occurs in those cells of a leaf that contain chloroplasts. The light-requiring reactions occur in the thylakoid and thylakoid membrane, and they produce ATP and NADPH. Excitation and transport of electrons take place in the thylakoid membrane. Hydrogen ions become concentrated in the interior of thylakoids.

ATP is produced by ATPases, which are embedded in the thylakoid membrane. In the stroma of the chloroplast, sugar is produced (Calvin cycle), using CO_2, existing sugars, energy from ATP, and electrons from NADPH.

Respiration. In animals and plants, most cells contain large numbers of mitochondria, which is the site of cellular respiration. In the inner mitochondrial compartment, sugars are broken down (Krebs cycle). As this happens, electrons and protons are stripped from sugar molecules, passed to carrier molecules, and then to an electron transport system localized in the inner mitochondrial membrane. Hydrogen ions become concentrated in the outer compartment (the intermembrane sac). ATP is produced by ATPases embedded in the inner mitochondrial membrane.

A big picture question, requiring the students to show that they know what happens where. The answers reproduced here reflect the main points of a good essay response. Given sufficient time, both of these essays might be elaborated on.

CHAPTER 20

UNIVERSITY OF COLORADO

EPOB 1220: GENERAL BIOLOGY II

Carol Ann Kearns, Associate Director, Environmental Residential Academic Program, and Senior Instructor of Environmental, Population, and Organismic Biology

FIVE EXAMS ARE GIVEN IN THIS COURSE. THE LOWEST GRADE IS DROPPED, OR THE student may opt out of the cumulative final if he or she takes all four exams. Only four exam grades are counted toward the course grade. An example of one of the four "midterms," plus the final, is included here. The intermediate exams are not cumulative in coverage, but the final is.

The primary objective of this course is to prepare students with a broad background of basic biological concepts. Students later taking more advanced courses refer back to their general biology coursework repeatedly as they begin to delve more deeply into specific topics. Our department offers another biology sequence, called "Biology, a Human Approach," for nonmajors who generally don't plan to take additional biology courses.

Key competencies and skills that should be developed as a result of taking this course include acquiring an adequate foundation of biological information to provide the tools to understand the biology they are confronted with in everyday life. In addition, students should have a strong foundation based on our current understanding of biology so that they are prepared for more advanced courses, and so that they can make the necessary links among different levels of biological thought. For example, those students who go on to take ecology should be able to look at ecological interactions with some notion of the physiological adaptations of organisms.

EXAM 1

Short-answer questions cover both factual material and conceptual material. Many of the questions require some integration of material that was presented in different contexts. Students should be able not only to recall information, but to integrate it and manipulate it in larger contexts.

Bryophyta
Psilophyta
Lycophyta
Sphenophyta
Cycadophyta
Ginkgophyta
Anthophyta
Coniferophyta
Pterophyta

Each phrase describes a phylum of plants. Choose the correct phylum from the list above. You may use a phylum to answer more than one question. There is only one correct phylum for each phrase.

Correctly answering this group of five questions requires recall of organismal diversity.

1. Plants with swimming sperm; gametophyte-dominant stage in life cycle.

Answer: Bryophyta

2. Plants with leaves that uncoil during development; spores produced in clusters on the lower surface of fronds.

Answer: Pterophyta

3. Plants that produce endosperm within the seed; most diverse group of plants on earth today.

Answer: Anthophyta

4. Scouring rushes or horsetails; plants with upright, green jointed stems and whorls of small leaves or branches at the joints.

Answer: Sphenophyta

5. Woody, cone-bearing trees; lower cones bear pollen and upper cones produce seeds.

Answer: Coniferophyta

The next four questions require recall of basic botany.

What part of the flower does each phrase describe?

6. Protects the flower bud; often photosynthetic.
Answer: sepals

7. Part of the stamen in which male gametophytes are produced.
Answer: anther

8. Often showy and colorful for pollinator attraction.
Answer: petals

haploid
diploid
triploid
dikaryotic

From the list of terms above, choose the term that best describes each of the structures.

The next five questions require an understanding of the significance of ploidy levels, the uniqueness of endosperm, the complexity of plant life cycles, and where specific structures occur during the life cycle.

9. endosperm
Answer: triploid

10. fern spore
Answer: haploid

11. liverwort sporophyte
Answer: diploid

12. pollen grain
Answer: haploid

13. sepal
Answer: diploid

14. Micronutrients generally function as

a. enzymes.
b. coenzymes.
c. major building blocks of organic molecules.
d. a source of amino acids.

Answer: b

15. Which best describes the role of potassium in plants?

a. component of ATP
b. essential for maintaining water balance
c. final electron acceptor in aerobic respiration
d. a component of the chlorophyll molecule

Answer: b

16. All of the following are true EXCEPT

a. Conifers and anthophytes both produce pollen.
b. Conifers produce seeds on cones and anthophytes produce seeds within a fruit.
c. Both conifers and anthophytes have very reduced gametophytes compared to mosses.
d. Both conifers and anthophytes produce endosperm.

Answer: d

17. Arrange these four events in the correct order to describe the flow of materials in the phloem:

w. Water diffuses into the sieve elements.
x. Leaf cells produce sugar by photosynthesis.
y. Solutes are actively transported into sieve elements.
z. Sugar moves through the phloem toward a sink.

Answer: x, y, w, z

Requires understanding the logic of transition and the functions of active transport and diffusion.

18. Which of the following is true?

a. The stele is a central cylinder composed entirely of ground tissue.
b. Monocot stems have a stele.
c. Dicot roots have a stele.
d. Root hairs are part of the stele.

Answer: c

19. Which of the following is most likely to occur when a small piece containing the primary meristem at the tip of a dicot shoot is cut off?

a. The plant will send out lateral branches.
b. The plant will lose its leaves.
c. The growth of buds at nodes is inhibited.
d. The plant grows tall and spindly.

Answer: a

Requires understanding of the role of the apical meristem in inhibition of lateral growth.

20. According to the pressure flow hypothesis of phloem transport,

a. sugars move from a high concentration source to an area that requires sugar.
b. water is actively transported into the source region of the phloem to create pressure.
c. evaporation of water from the leaves creates a tension that pulls sugars through the phloem.
d. the companion cells act as pipelines for the transport of sugars.

Answer: a

The question concerns the mechanism of translocation and its distinction from transpiration.

21. The opening of stomates is thought to result from

a. an increase in the concentration of K^+ ions in the guard cells.
b. active transport of water into the guard cells.
c. decreased water in the guard cells.
d. the buildup of too much CO_2 gas within the leaf.

Answer: a

22. Water and dissolved particles that travel along the cell walls of root cells travel through the

a. apoplast.
b. symplast.
c. plasmodesmata.
d. phloem.

Answer: a

23. Which of the following is NOT one of the advantages of having an internal fluid-filled cavity such as a coelom or pseudocoelom?

a. allows simple circulation

b. creates a gastrovascular cavity for the digestion of food

c. cushions the organs

d. serves as a site for gamete maturation

Answer: b

24. Which of the following exhibits radial symmetry?

a. medusa

b. tapeworm

c. trematode

d. sponge

Answer: a

This requires recall of diversity as well as recall of symmetry types and adaptive values.

25. Which of the following are not found in sponges?

a. spicules

b. flagella

c. cellular level of organization

d. nematocysts

Answer: d

Fill in the blank:

The next six questions require relating morphological form to function.

26. The male gametophyte of the Anthophyta is called ______.

Answer: pollen

27. Pollen tubes must grow down the ______ to reach the ovules, an elongate portion of the female reproductive structure.

Answer: style

28. A ripened ovary is called a ______.

Answer: fruit

29. An irregularly shaped support cell lacking cytoplasm and having lignified cell walls is called a ______ .

Answer: sclereid

30. The attraction that holds water molecules to other water molecules is called ______.

Answer: cohesion

31. A blastula folds in on itself to produce a two-layered cup-shaped stage called a ______.

Answer: gastrula

Corrections

Correct the following statements. Do NOT correct them by merely making a negative statement such as this:

Statement: The dog said meow.
Response: WRONG—The dog did NOT say meow. CORRECT—The CAT said meow.

Correction of each sentence requires justification and minimizes guessing.

32. An increase in plant height results from division of cells of the pericycle.
Answer: "apical meristem" replaces "pericycle"

33. Companion cells are wide, short, water-conducting cells which are stacked end-to-end to form continuous tubes.
Answer: "sieve tube elements" replaces "companion cells"

34. In a tree the young, active xylem and phloem are located in the heartwood.
Answer: "sapwood" replaces "heartwood"

35. The comblike rows of cilia on the ctenophores are known as tentacles.
Answer: "ctenes" replaces "tentacles"

36. Muscle tissues are derived from ectoderm.
Answer: "mesoderm" replaces "ectoderm"

37. Flowers that are wind-pollinated will probably be fragrant and colorful and have nectar guides.

Answer: "bee-pollinated" replaces "wind-pollinated"

Other valid possibilities also exist for this question.

38. List three factors that help determine whether stomates will be open or closed.

The question reviews the function of stomates, why they are important, and the logic of why they should be open or closed. Also incorporated in the question are concepts associated with transpiration, gas exchange in photosynthesis, osmosis, active transport, turgor pressure, and the relationship between form and function. To answer this question successfully, students should:

1. Think about the function of stomates.
2. Based on their function, decide when and why stomates should be open.
3. Determine what factors will stimulate stomates to be open under these specific conditions.
4. Determine the function of closed stomates.
5. Describe the conditions under which stomates should close.
6. Describe a mechanism for closure.

Answer:

1. Circadian rhythms: Stomates are generally open in daylight when gas exchange is needed for photosynthesis.

2. Light: Photosynthesis generates ATP, which fuels the K^+ pump that pumps ions into guard cells. Water follows by osmosis, and guard cells become turgid. Guard cells open due to orientation of fibers in cells. Decreased CO_2 concentration in the leaf during photosynthesis stimulates opening. CO_2 is needed for photosynthesis.

3. Stomates close under conditions when plants are losing too much water through transpiration. The balance between photosynthesis and transpiration temperature and water balance can override other factors involved in opening of stomates. Loss of water in guard cells decreases turgidity. Because of orientation of fibers in guard cells, decreased turgidity leads to flaccidity and closure of stomates.

39. What are the advantages of sexual and asexual reproduction in plants? What are the disadvantages of each?

This is a critical question. At this point, students have already studied mitosis and meiosis. Now these concepts are applied to plants. Students must

1. Realize the nature of gene shuffling that occurs in meiosis.
2. Understand the evolutionary significance of genetic variation.

3. Understand the short-term conditions when offspring survival is enhanced by genetic variation.
4. Understand the investment and risk associated with sexual reproduction in plants.
5. Understand the importance of asexual reproduction.

Answer: Sexual reproduction produces genetic variation under changing environmental conditions or with dispersal to new environments. The variation is likely to result in some offspring that will thrive in their environments. Genetic variation is critical for long-term survival.

Sexual reproduction may involve a high energy investment in fruit or seeds or floral displays. Under some conditions, pollination or fertilization levels may be inadequate. If parent plants are thriving and highly adapted to an environment, varied offspring may not be as well suited to that site.

Asexual reproduction produces genetically identical offspring, which have a good chance of surviving in the parental habitat if parents are well suited to that environment. Asexual reproduction may not be as costly as sexual reproduction since it is not always necessary to invest in specialized reproductive structures. Asexual reproduction may simply take the form of rhizomatous growth.

Asexual reproduction does not produce genetic variation or novel combinations of genes. Offspring may not be suited to changing environments, and long-term adaptation is limited.

40. Compare the way in which fertilization takes place in the mosses and in the Anthophyta.
How do the gametophytes differ?
How do the gametes differ?
How do the male gametes get to female gametes?

This is a major question, which draws on concepts of plant life cycles, ploidy levels, increasingly terrestrial adaptations of more recently evolved plant divisions, and pollination biology.

Answer: Mosses have independent photosynthetic gametophytes. The haploid gametophytes are the dominant stage of the life cycle. They produce stationary female gametes and swimming sperm. A minimum of a film of water is essential for the sperm to swim to the egg cells to effect fertilization.

Anthophytes have dependent, highly reduced female gametophytes found within the ovules, within the ovaries of flowers. Male gametophytes are pollen grains consisting of only a few haploid nuclei. Both sets of gametophytes are produced on the sporophyte plant. The pollen grains must generally be transported by vectors (insects, birds, bats, wind, etc.) to effect pollination upon arrival at stigmas. For fertilization to occur, a pollen tube must grow down the style, and two male nuclei must migrate down the pollen tube. Fertilization involves the union of one male nucleus with an egg nucleus to produce a zygote, and one male nucleus with two haploid female nuclei to produce endosperm.

FINAL EXAM

Students who have the most difficulty with the General Biology sequence are those who fail to stay on top of the material. We cover a tremendous amount of information, and the student who does not read the text and go over some part of the material three or four times a week is generally unable to catch up before an exam.

Reread class notes from the previous class immediately before attending class. Do not fall behind on assigned readings. Rewrite (or type out) class notes as a means of reviewing the material. Go over all the class notes for the unit repeatedly. Three or more times a week, spend an hour reviewing the materials covered in the unit. Do this in addition to keeping up with assigned readings.

1. The archaebacteria are sometimes placed in their own domain, rather than in the Kingdom Monera. The main difference between the eubacteria and the archaebacteria is

a. staining properties.
b. size.
c. differences in rRNA and other essential macromolecules.
d. that only archaebacteria can photosynthesize.

Answer: c

2. The endosymbiotic model suggests that

a. muscle contraction results from the shortening of actin filaments.
b. a coelom can serve as a reservoir for wastes.
c. cell organelles developed through the infolding of the plasma membrane.
d. eukaryotic cells are derived from a collections of protists living in close association.

Answer: d

3. Where would you be most likely to find diatoms (Phylum Bacillariophyta)?

a. the canopy of the tropical rainforest
b. the soil of the savanna
c. in the ocean, often in association with coral reefs
d. in the conifers of the taiga

Answer: c

4. In fungi, karyogamy does not immediately follow plasmogamy, which

a. means that sexual reproduction can occur in specialized structures.
b. results in more genetic variation during sexual reproduction.
c. allows fungi to reproduce asexually most of the time.
d. creates dikaryotic cells that may benefit from the presence of duplicate copies of alleles.

Answer: d

5. The symbiotic associations called mycorrhizae are considered to be

a. parasitic.
b. mutualistic.
c. the beginning stages of the formation of lichens.
d. strictly asexual.

Answer: b

6. Bryophytes have all of the following characteristics EXCEPT

a. multicellularity.
b. well-developed vascular tissue.
c. a protected, stationary egg cell.
d. a reduced, dependent sporophyte.

Answer: b

7. Along with the seed, the seed plants have evolved several additional adaptations to the terrestrial environment. Which one of the following is NOT such an adaptation?

a. Flagellate sperm swim to the ovule to effect fertilization.
b. The female gametophyte is protected from desiccation by the surrounding tissues of the sporophyte.
c. The seed and/or associated structures serve as a means of dispersal.
d. Pollen serves as the male gametophye.

Answer: a

8. All of the following could be considered advantages of asexual reproduction in plants EXCEPT

a. success in a stable environment.
b. increased agricultural productivity.
c. increased ability to adapt to change.
d. ability to clone an exceptional plant.

Answer: c

9. The following are all true statements about the structure of roots and shoots EXCEPT

a. Only shoots have nodes.
b. Only shoots have endodermis.
c. Lateral shoots arise from leaf axils.
d. Roots lack stomata.

Answer: b

10. Which of the following best describes the general role of micronutrients in plants?

a. They are cofactors in enzyme reactions.
b. They are necessary for essential regulatory functions.
c. They are components of nucleic acids.
d. They are components of cell walls.

Answer: a

11. In a typical multicellular animal, the circulatory system interacts with specialized surfaces in order to exchange materials with the exterior environment. Which of the following is NOT an example of such an exchange surface?

a. lung
b. muscle
c. skin
d. kidney

Answer: b

12. In what type of organism might you find an ascus?

a. nematode
b. fungus
c. ciliate
d. Ginkgophyta

Answer: b

13. What are essential amino acids?

a. those that are absent in fruits and vegetables
b. the only amino acids found in human proteins
c. one class of vitamins that is indispensable for neurological development
d. molecules obtained from food in a prefabricated form because they can't be synthesized by the animal

Answer: d

14. Which tissue type does this describe? Functions as a covering and lining; serves as a barrier against mechanical injury, fluid loss, and microbial invasion; cells are tightly joined together; often secretory tissue.

a. connective

b. smooth muscle

c. epithelial

d. nervous

Answer: c

15. The resting potential of a nerve cell is about –70 mV. This means that

a. the interior of the cell is negative with respect to the exterior.

b. the extracellular environment is negative with respect to the interior of the cell.

c. the cell is depolarized.

d. the sodium concentration is greater inside the cell than outside the cell.

Answer: a

16. The hormone calcitonin decreases blood calcium concentration. The hormone parathormone increases blood calcium concentration. What is the advantage of having two hormones regulating blood calcium concentration?

a. Two opposing hormones are essential for negative feedback mechanisms to work.

b. Antagonistic hormones allow strict control of blood calcium levels.

c. This permits blood calcium levels to change dramatically after a meal.

d. There is no advantage.

Answer: b

17. Which of the following is most likely to describe a body surface specialized for gas exchange?

a. highly vascularized, moist, membranous tissue with a large surface area

b. smooth muscle layer lined internally with epithelial tissue; the epithelial tissue secretes hydrolytic enzymes

c. excitable tissue with specialized connections between cells, which allow rapid conduction of impulses

d. specialized system of tubules functioning in filtration, reabsorption, and secretion; tubes intimately associated with capillary bed

Answer: a

18. Which of the following is NOT directly dependent on a person's adequate protein intake?

a. formation of enzymes
b. formation of glycogen stores in the liver
c. production of antibodies in the immune system
d. production of muscle tissue

Answer: b

19. Which of the following is NOT a function of the vertebrate liver?

a. converts excess glucose to glycogen
b. deaminates excess amino acids
c. produces bile
d. conserves body water

Answer: d

20. Which of the following structures is INCORRECTLY matched to its function?

a. epididymis—maturation and storage of sperm
b. fallopian tube—site of normal embryonic implantation
c. prostate gland—adds alkaline substances to semen
d. testes—production of testosterone

Answer: b

21. In general, valves in the circulatory system

a. permit blood to circulate rapidly.
b. prevent blood from moving too rapidly.
c. prevent blood from flowing in the wrong direction.
d. stop the circulation whenever necessary.

Answer: c

22. Hormones

a. are released from the ducts of glands directly onto target cells.
b. travel in the bloodstream to cells that may be quite some distance from where the hormone is made.
c. should be banned from college campuses.
d. are the same as neurotransmitters.

Answer: b

23. Excess secretion from this gland can cause a person to be thin, hyperactive, always hungry and irritable, with a high metabolic rate.

a. adrenal cortex
b. thyroid

c. pancreas
d. thymus

Answer: b

24. A neurotransmitter

a. is a liquid that travels from one nerve cell to the next through a central hollow tube that connects the two cells.
b. forms the fluid cushion around the brain and spinal column.
c. is a substance released at the end of one nerve that sends a signal to the next cell.
d. is an electrical impulse that jumps from one node to the next along the myelin sheath that covers a nerve cell.

Answer: c

25. Lymph nodes

a. filter microbes out of a fluid that eventually enters the blood.
b. produce red blood cells.
c. are attracted to damaged tissue by a process known as chemotaxis.
d. filter microbes directly out of the blood.

Answer: a

26. Which statement follows from the principle of allocation?

a. The number of organisms an area can support is determined by its energy supply.
b. Physiological adjustments to environmental changes can extend the abilities of organisms to adjust to stress.
c. The total amount of energy available to an organism is partitioned into such processes as reproduction, growth, and coping with environmental stresses.
d. Organisms that allocate most of their energy to reproduction are better at surviving in variable environments than other organisms.

Answer: c

27. In general, deserts are located at latitudes where

a. dry air is descending.
b. moist air is descending.
c. dry air is rising.
d. air masses are stationary.

Answer: a

28. All of the following statements about the logistic model of population growth are correct EXCEPT

a. it incorporates the concept of carrying capacity.
b. it describes differences in population growth rates over time.
c. it describes a period of exponential growth when the population size is very small.
d. it predicts population size will stabilize when birth rates exceed death rates.

Answer: d

29. Which is a feature of amphibians?

a. endothermic
b. amniotic egg
c. respiration through skin
d. open circulatory system

Answer: c

30. Which one of the following phyla does not exhibit the same basic symmetry as the others?

a. Platyhelminthes
b. Ctenophora
c. Annelida
d. Nematoda

Answer: b

31. All of these phrases describe a tropical rain forest, EXCEPT

a. nutrient-rich soil.
b. many arboreal plants and animals.
c. great diversity of tree species.
d. little light filtering to forest floor.

Answer: a

32. Which statement about hormones is correct?

a. Hormone effects are local, felt only in the area of the gland that produces the hormone.
b. A hormone acts directly on all cells of an organism.
c. A hormone changes the physiological state of target cells bearing receptors specific to that hormone.
d. Hormones are transported through the body from one cell to the next by the process of diffusion.

Answer: c

33. Which is the division of the peripheral nervous system associated with involuntary responses?
    a. spinal cord
    b. somatic
    c. autonomic
    d. brain

Answer: c

34. The following diagram represents the trophic levels of an ecological pyramid (energy pyramid).

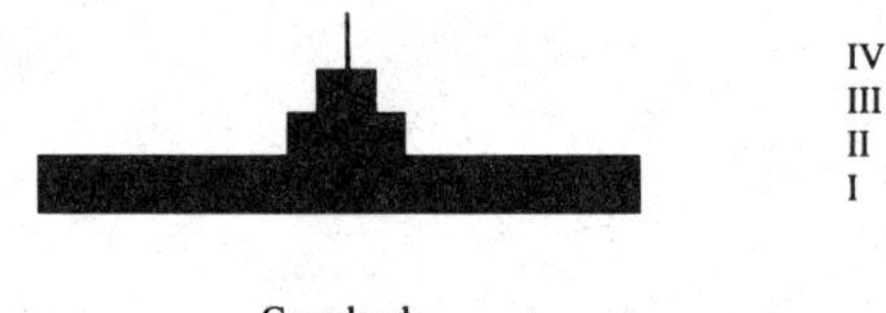

What does "I" represent?
    a. primary producers
    b. primary consumers
    c. secondary consumers
    d. tertiary consumers

Answer: a

35. Which one of the following taxa is composed of animals that lack jaws?
    a. Chondrichthyes
    b. Osteichtheyse
    c. Agnatha
    d. Amphibia

Answer: c

36. Explain how each of the following factors could have different effects on human population growth in different parts of the world.
    a. generation time
    b. age structure

We spend a considerable amount of time studying population growth, and we apply this to human populations and the world population crisis. Students should be able to personalize some of the information to help them make reproductive choices in the future. The answer to this question requires an understanding of basic demographic concepts.

Answer: Shorter generation times increase the rate of population growth. A population in which females bear 3 individuals starting at age 20 grows more rapidly than a population in which females bear 5 children starting at age 30.

Students should be able to diagram age-structured populations for stable, decreasing, and growing populations. They must understand the concept of population momentum, and why populations continue to grow even after the average fertility rate drops. They should be able to relate this information to the stages of demographic transitions.

Answer (continued): Age structure affects the rate of population growth. The more individuals there are who are in and below their reproductive years, the more rapidly a population grows. The age-structured population diagram of a rapidly growing population looks like a pyramid. A stable population has roughly equal numbers of individuals in all age categories (except the oldest).

37. Describe transpiration.

Surprisingly, by the end of the year, students still confuse the terms *transpiration, translocation,* and *transduction.*

1. **Students must recall the correct concept!**
2. **Students must describe all the forces involved in transpiration.**
3. **Students must recall the form of the structures involved in transpiration.**
4. **Students must know the source of energy for transpiration.**
5. **Students must know the magnitude of transpiration compared to other forces, such as root pressure, which move water in plants.**

Answer: Transpiration is the solar-powered movement of water through plants. Water enters plant roots due to active transport of ions followed by osmosis. Water enters into the apoplast of the stele and is transported up xylem in tracheids or vessels. Tracheids and vessels are the remains of once-living cells whose cytoplasm has died, leaving hollow cylinders formed of cell walls.

Stomates are openings on the surface of leaves that permit gas exchange between the leaf and the atmosphere. When stomates are open, and the sun is shining, water evaporates from the stomates at a high rate. The constant removal of water through the stomates pulls water up through the xylem. Adhesion of polar water molecules to the walls of the xylem, and cohesion of polar water molecules to each other results in the movement of columns of water up the xylem tubes.

38. What is the difference between a coelom and a pseudocoelom?

In studying animal diversity, we classify animals by level of organization (tissue, organ system), and by developmental characters related to the presence or absence of internal body cavities. Students should know

1. Why we stress this developmental stage in classification.
2. The advantages of having an internal fluid-filled cavity.
3. The advantages of a coelom over a pseudocoelom.
4. Students should make the connection between the fact that coeloms are lined by mesoderm, and that mesoderm can develop into muscle tissue, allowing organisms to develop two independent muscle systems.

Answer: Coeloms and pseudocoeloms are internal fluid-filled cavities found in animals. Both structures can serve as primitive storage areas, primitive circulatory systems, can cushion organs, and can serve as a hydrostatic skeleton. Pseudocoeloms are surrounded on one side by mesoderm and on one side by endoderm. Coeloms are surrounded by mesoderm on both sides. Muscle tissue is derived from mesoderm. This means that the organism can have muscle tissue associated with both the outer body wall and the gut. Therefore, the organism can perform digestive activities (peristalsis) independently of body movement. In addition, the mesoderm forms mesentaries, which anchor organs in place. The true coelom is considered a more advanced form. Mammals are an example of a taxon of coelomate animals. Rotifers are an example of pseudocoelomates.

39. What is a keystone predator?

This is a key ecological concept. Students should be able to relate the idea of predation to community structure. They should be able to explain how one species can affect species diversity.

Answer: A keystone predator is an organism whose feeding preferences are critical to maintaining species diversity in a community. A keystone predator generally eats a highly competitive species that is lower on the food chain. By keeping populations of the competitive species in check, other species can coexist in the habitat. Populations of these other species will be suppressed or eliminated due to competition if the predator is eliminated. An example of a keystone predator is a starfish that lives off the coast of California. This starfish eats mussels, which tend to outcompete other invertebrates. If the starfish is eliminated, the mussels dominate the community at the expense of other invertebrate species. Thus the starfish maintains species diversity by holding the mussel population in check.

40. What are the four key features that characterize chordates?

In covering diversity, we stress that the majority of animals are invertebrates. Most students are much more familiar with the vertebrate groups. Therefore, when we reach the unit on chordates, we make it clear that not all chordates are vertebrates. Even within their own phylum, vertebrates are joined with invertebrates. The reason these invertebrates are grouped with the vertebrates is because of the presence of four key features present at some stage of the life cycle of all these organisms. We go on to stress the evolutionary development of these structures and how their functions change from one group to the next. Thus, although this question only asks students to recall the features themselves, they have been exposed to the features in many contexts.

Answer:
pharyngeal gill slits
notocord
dorsal hollow nerve cord
post-anal tail

41. Compare plants and fungi. Give at least four differences between the two kingdoms. (Be sure to give the corresponding information for each kingdom—e.g., plants have ______, but fungi have ______.)

Much of the semester is spent covering diversity issues. We study plants and fungi independently. This question asks students to take two separate units and to integrate them.

Answer: Plants have chlorophyll and are autotrophic (with some highly evolved parasitic exceptions), and fungi are heterotrophic, absorbing small predigested molecules.

Plants have a life cycle involving alternation of a haploid gametophyte stage and a diploid sporophyte stage. Fungal mycelia are either haploid or dikaryotic (bearing two genetically distinct nuclei).

Plants have male and female gametes. Some fungi display male and female structures, but many have mating strains (+, –) that are morphologically similar.

Plants have cellulose cell walls, and fungi have chitinous cell walls.

Some plants have flagellate male gametes. Flagella are absent in the Kingdom Fungi.

Note that there are many other possible answers to this question.

CHAPTER 21

UNIVERSITY OF LOUISVILLE

BIOLOGY 240: DIVERSITY OF ORGANISMS

Peter Sherman, Assistant Professor

FOUR EXAMS ARE GIVEN IN THE COURSE, THREE "MIDTERMS" AND A FINAL. AN EXAMPLE of a midterm and final are included here. All are multiple-choice exams, which, together, account for 100 percent of the grade.

The main objective of this course is to introduce students to the concepts of evolution and to provide them with a survey of the biological diversity of living organisms. To achieve this, I want my students to gain an understanding of how natural selection has brought about a variety of different solutions to life in different environments; how simple body plans have been modified over time into more complex organisms (unicellular to multicellular, development of organ systems in animals, etc.); to understand that evolution depends on chance mutations and results in adaptations to the present environment, but is not progressive, resulting in the production of better and better organisms. I also want them to gain an appreciation for the vast diversity of living things that we share the planet with, and gain an understanding of the basics of taxonomy and systematics, including how systematists work out phylogenies.

Course exams test knowledge of the sequence of evolutionary events, how systematists have decided to group different sets of organisms into common phyla, how different organisms adapt to their particular environments, what unique adaptations different taxa of organisms exhibit, concepts of evolution and how it occurs, and so on.

This is a good deal of material; therefore, spend time learning the material as we go along rather than trying to cram it all in right before the exam. Also, outline your notes to help speed up reviewing for exams. As you read through the outline, you can skim past material that you remember and go back to your notes for material you can't fully recall. You might make charts that list the different groups of organisms by phylum and indicate what characteristics they have or don't have—for example, for animals: type of symmetry, number of germinative layers, level of organization (cellular, tissue level, organ systems), body cavity and type, if present, and so on. Be sure to read the textbook to help reinforce material covered in lecture.

MIDTERM EXAM

1. In the five-kingdom classification of organisms, the five kingdoms are
a. Protozoa, Monera, Animalia, Plantae, Fungi.
b. Protozoa, Protista, Animalia, Plantae, Fungi.
c. Prokaryota, Protozoa, Animalia, Plantae, Fungi.
d. Prokaryota, Protista, Animalia, Plantae, Fungi.
e. none of the above

Answer: e

2. DNA in the nucleus is duplicated during interphase.
a. true
b. false

Answer: a

3. You might not want to sit next to someone who had an abnormally high level of flatus because
a. You might catch the bubonic plague from them.
b. You might catch tuberculosis from them.
c. You might catch pneumonia from them.
d. They might be making some really bad smells.

Answer: d

4. All living things are composed of one or more cells.
a. true
b. false

Answer: a

This tests understanding of a major point. Students often think that the human body plan has the same basic elements as all other organisms.

5. **The organisms that are found in 3.8-billion-year-old fossils are**
   a. Fungi.
   b. Protista.
   c. Plantae.
   d. Monera.
   e. none of the above

Answer: d

This tests mastery of factual material as well as understanding of the evolutionary progress of life on earth.

6. **What is the exact name of the main primary pigment in all plantlike protista?**
   a. chlorophyll a
   b. chlorophyll b
   c. chlorophyll c
   d. chlorophyll d

Answer: a

7. **All living things exhibit metabolism, complex organization, growth, and respond to stimuli.**
   a. true
   b. false

Answer: a

Requires understanding of a major point—that is, the characteristics that biologists use to define what is living versus what is nonliving.

8. **The organism that causes the disease malaria belongs to what phylum?**
   a. Rhizopoda
   b. Dinoflagellata
   c. Apicomplexa
   d. Bacillariophyta

Answer: c

9. **The cell wall of archaebacteria does not contain peptidoglycan.**
   a. true
   b. false

Answer: a

A test of mastery of factual material, the question also touches on one of the key traits used to separate the two major bacteria groups, archaebacteria and eubacteria.

10. In the slime mold life cycle, the sluglike mass composed of many individual slime mold cells forms when
   a. predators become more common.
   b. food or water becomes scarce.
   c. it is time for the slime mold to dig a hole in the ground to hibernate for the winter.
   d. it is time for the slime mold to migrate to a warmer location for the winter.

Answer: b

11. During alternation of generations in seaweeds, spores give rise to ______.
   a. gametes
   b. gametophytes
   c. sporophytes
   d. sporozooids

Answer: b

This tests mastery of factual material and understanding of a major concept. Students often have difficulty understanding alternation of generations in plants. The key to getting it straight is to recall that sporophytes produce spores, while gametophytes produce gametes. The generations alternate because sporophytes have gametophyte babies and gametophytes have sporophyte babies; thus spores (produced by sporophytes) give rise to gametophytes.

12. What is the name of the structure that detoxifies alcohol in the liver?
   a. smooth endoplasmic reticulum
   b. rough endoplasmic reticulum
   c. Golgi apparatus
   d. nucleus
   e. mitochondrion

Answer: a

13. A group of very closely related species that share a number of similar traits is called a (an)
   a. phylum.
   b. order.
   c. genus.

d. class.
e. kingdom.

Answer: c

Requires understanding of a major principle: that taxa are arranged from broad categories into increasingly narrow categories; thus the next largest grouping that would encompass a group of closely related species would be genus. A mnemonic for remembering the taxonomic categories is: *Kings Put Class on Families Going Somewhere* (kings can endow a family with nobility). This gives you the correct descending order for Kingdom, Phylum, Class, Order, Family, Genus, Species.

14. Recent evidence suggests that these two groups of organisms, which traditionally have been grouped together in the same kingdom, should be separated into two different kingdoms. The groups are
a. prokaryotes and eukaryotes.
b. algae and protozoans.
c. archaebacteria and eubacteria.
d. unicellular algae and seaweeds.

Answer: c

A correct answer requires understanding of how phylogenetic groupings are created.

15. What is the name of the group of archaebacteria that live in very salty environments?
a. thermoacidophiles
b. extreme halophiles
c. extreme saliphiles
d. gram-positive bacteria

Answer: b

16. In the ciliates, the micronucleus contains DNA that is
a. exchanged with other ciliates during conjugation.
b. used to make proteins needed by the ciliate.
c. used to produce either haploid sperm or haploid eggs.
d. used to produce haploid spores.

Answer: a

17. The reproductive cells involved in asexual reproduction in seaweeds that undergo alternation of generations are called

a. spores.
b. sperm.
c. gametes.
d. eggs.

Answer: a

In plants, gametes are haploid and involved in sexual reproduction, spores are diploid and involved in asexual reproduction.

18. Mitochondria are the
a. main site where fatty compounds are synthesized.
b. main site where ATP is produced.
c. main site where DNA is stored.
d. main site where proteins are synthesized.

Answer: b

19. The Phylum Rhizopoda
a. contains organisms that have flagella.
b. contains organisms that have pseudopods.
c. contains organisms that have cilia.
d. all of the above
e. none of the above

Answer: b

20. Sharks and dolphins are similar in appearance because of ______.
a. convergent evolution
b. common ancestry
c. sexual reproduction
d. asexual reproduction

Answer: a

A correct response requires understanding of a major principle of biology. The question touches on the need for systematists to separate out similarities due to common ancestry from similarities due to convergent evolution.

21. Systematics is the
a. branch of biology concerned with evolutionary relationships among organisms.

b. branch of biology concerned with the scientific classification of organisms.
c. branch of biology concerned with the diversity of life.
d. branch of biology concerned with the study of multicellular organisms.

Answer: a

22. A plasmid is

a. an organelle that occurs in plants and algae.
b. a small ring of DNA in a bacteria.
c. an organelle that occurs in plant and animal cells.
d. a type of bacteria that lives in the rumen of cows.

Answer: b

***Plasmids* are found in bacteria, and *plastids* (e.g., chloroplasts) in plants.**

23. The bacterial species that appears to be the major cause of tooth decay is

a. *Clostridium tetani.*
b. *Yersinia pestis.*
c. *Streptococcus mutans.*
d. *Plasmodium vivax.*

Answer: c

24. Trypanosoma is an organism that causes

a. malaria.
b. bleeding sores that won't heal and dissolving of mucous membranes.
c. sleeping sickness.
d. leprosy.

Answer: c

25. The antibiotic known as penicillin

a. interferes with the synthesis of peptidoglycan.
b. interferes with the synthesis of the bacterial cell membrane.
c. prevents the bacterial ribosomes from functioning normally.
d. prevents the bacterial DNA from functioning normally.

Answer: a

26. Cellular division that produces daughter cells with half the number of chromosomes of parent cells is called

a. binary fission.

b. mitosis.
c. meiosis.
d. conjugation.
e. cytokinesis.

Answer: c

Requires understanding of a very important biological principle: mitosis conserves chromosome number, meiosis halves it.

27. Gram-positive bacteria

a. do not have an outer membrane.
b. have very little peptidoglycan in their cell wall.
c. are stained red by the Gram's stain.
d. More than one of the above statements is true.

Answer: a

28. The parasites that cause the disease malaria reproduce in humans inside

a. the intestine.
b. the skin.
c. the fluid surrounding the brain.
d. red blood cells.

Answer: d

29. What do both mitochondria and chloroplasts have in common?

a. They are structures where ATP is produced.
b. Both are found in all eukaryotic cells.
c. Both may be found in some prokaryotic cells.
d. Only b and c are correct.
e. a, b, and c are correct.

Answer: a

Calls for understanding basics of cell structure.

30. One of the symptoms of the disease tetanus is

a. diarrhea.
b. continuous muscle relaxation.
c. continuous muscle contractions.

d. constipation.

Answer: c

31. Which of the following are in the correct order?

a. Kingdom, Phylum, Class, Family, Order, Genus, Species
b. Kingdom, Phylum, Order, Class, Family, Genus, Species
c. Kingdom, Phylum, Class, Order, Family, Genus, Species
d. Kingdom, Phylum, Class, Order, Genus, Family, Species

Answer: c

See question 13.

32. Dr. Sherman (the person who teaches this class) has a scar caused by leishmaniasis on his

a. arm.
b. buttocks.
c. leg.
d. back.

Answer: c

This tests whether or not students have been attending class.

33. The Golgi apparatus is

a. the main site where fatty compounds are synthesized.
b. the main site where ATP is produced.
c. involved in sorting and packaging lipids and proteins.
d. the main site where proteins are synthesized.

Answer: c

Requires recall of factual material concerning cell function.

34. Which of the following would most likely occur if all members of the group Monera were suddenly and permanently to disappear from the face of the earth?

a. Animal populations would eventually increase due to the absence of disease.
b. All life would eventually die due to the decrease in the number of organisms that decompose dead organic matter.
c. Only the organisms that feed on monerans would decrease in number.

d. Very little change would occur.

Answer: b

Contrary to popular belief, harmful bacteria are in the minority. The vast majority of bacteria are decomposers, which help (for example) to maintain the nutrient cycle that returns useful substances to the soil through their breakdown of dead organic matter.

35. The first fossils of eukaryotic cells show up in rocks that are about ____ billion years old.

a. 1.5
b. 2
c. 2.5
d. 3
e. 3.8

Answer: a

36. Metabolism is the name given to the chemical processes that occur in living cells.

a. true
b. false

Answer: a

37. A group of interbreeding individuals that is reproductively isolated from other groups is called a(n)

a. phylum.
b. organism.
c. family.
d. species.
e. genus.

Answer: d

Calls for understanding a fundamental principle: species is the only taxonomic grouping that can be biologically determined. All higher taxonomic groupings are subjectively determined, using the best available evidence.

38. Sex pili are used during

a. binary fission.
b. mitosis.
c. ruciosis.

d. conjugation.
e. cytokinesis.

Answer: d

39. Red tides are caused by members of the phylum

a. Phaeophyta.
b. Bacillariophyta.
c. Rhodophyta.
d. Chlorophyta.
e. Dinoflagellata.

Answer: e

40. Exotoxins are

a. toxic molecules found mostly in the cell walls of gram-negative bacteria.
b. toxic molecules found mostly in the cell walls of gram-positive bacteria.
c. toxic waste products produced mainly by gram-negative bacteria.
d. toxic waste products produced mainly by gram-positive bacteria.

Answer: d

41. Choose the one statement listed below that is FALSE:

a. The organism that causes amebic dysentery is in the Phylum Apicomplexa.
b. The genus for domestic cats and for other wild species of small cats is *Felis.*
c. The surface of the rough endoplasmic reticulum is rough due to the presence of smaller organelles called ribosomes.
d. Mitosis results in daughter cells that contain the same number of chromosomes as the parent cell.

Answer: a

42. Which of the following groups make up an important part of the phytoplankton?

a. Phacophyta
b. Foraminifera
c. Rhodophyta
d. Dinoflagellata
e. none of the above

Answer: d

43. What is the name of the hairlike appendages that help bacteria adhere to surfaces?

a. flagella
b. pili

c. cilia
d. capsule

Answer: b

Requires knowledge of bacterial structures.

44. The genus of the bacteria that causes botulism is

a. Clostridium.
b. E. coli.
c. Botulinum.
d. Yersinia.
e. Pestis.

Answer: a

Not only requires recall of factual material, but the ability to recognize a genus versus species.

45. The organisms that live in oceans and glow when they are touched or disturbed are in the phylum

a. Bacillariophyta.
b. Phacophyta.
c. Rhodophyta.
d. none of the above

Answer: d

46. Which of the following is NOT a protein?

a. collagen
b. ATP
c. growth hormone
d. antibodies

Answer: b

Requires basic knowledge about proteins.

47. In the slime mold life cycle, the sluglike mass composed of many individual slime mold cells crawls around until it finds an area that is

a. extremely moist.
b. exposed to sunlight.

c. in the shade.
d. extremely dry.

Answer: b

Guesswork will lead to answer *a*, but recall of lecture will yield *b*, the correct response.

48. Cilia are a characteristic of organisms in the phylum

a. Ciliophora.
b. Rhizopoda.
c. Zoomastigophora.
d. Apicomplexa.
e. none of the above

Answer: a

Mastery of fact is helpful, but so is a common-sense reading of the prefixes.

49. Which of the following would be found in an animal cell, but not in a bacterial cell?

a. DNA
b. ribosomes
c. cell wall
d. endoplasmic reticulum
e. cell membrane

Answer: d

Tests understanding of some of the key similarities and differences between prokaryotes and eukaryotes.

50. In multicellular seaweeds, the gametophyte

a. is haploid.
b. produces haploid reproductive cells.
c. produces diploid reproductive cells.
d. a and b
e. a and c

Answer: d

The gametophyte is haploid and produces haploid gametes, which are used in sexual reproduction. Cells used for sexual reproduction must be haploid to avoid doubling the number of chromosomes

in the next generation. In contrast, sporophytes are diploid and produce diploid reproductive cells called spores. Spores are involved in asexual reproduction. If they were used for sexual reproduction, the offspring produced would have twice as many chromosomes as their parents.

FINAL EXAM

This may be helpful: After you feel you have learned your notes fairly well, get together in a small group with other students who have studied their notes and quiz each other. As the course progresses, students should begin to get a feel for the type of questions the professor is likely to ask, and they can quiz each other by trying to anticipate what is likely to be asked on the upcoming exam.

1. What is the SUBPHYLUM (not phylum) of the chordate that surrounds its body with a protective covering made of cellulose?

a. Hemichordata
b. Cephalochordata
c. Urochordata
d. Chordata

Answer: c

2. A person who is infected with HIV, but who has not developed AIDS, cannot infect other people with HIV.

a. true
b. false

Answer: b

Tests factual knowledge as well as understanding of how viruses are transmitted from one host to another.

3. People who have the disease sickle cell anemia will have the following genes:

a. HH.
b. Hh.
c. hh.
d. Two of the above people will have the disease sickle cell anemia.

Answer: c

A correct response requires an understanding of how dominant and recessive genes work.

4. Every nucleotide contains

a. amino acids and RNA.

b. a sugar, a phosphate group, and a nitrogenous base.

c. amino acids and protein.

d. adenine, guanine, and cytosine.

Answer: b

5. Which of the following are made of keratin?

a. bird feathers

b. reptile scales

c. human hair

d. two of the above

e. all of the above

Answer: e

> **This tests factual material as well as understanding of the evolutionary concept that mammals and birds evolved from reptiles.**

6. Speciation often occurs in animals when

a. two different species interbreed and produce offspring different from either parent.

b. the zygote undergoes a mutation during development.

c. a geographical barrier separates two populations of the same species for a very long period of time.

d. a species lives in an environment where it is exposed to direct sunlight for long periods of time.

Answer: c

> **A correct response requires knowledge of process of evolution.**

7. Evolutionary biologists think that the group ______ evolved from reptiles.

a. Aves

b. Amphibia

c. Mammalia

d. two of the above

e. all of the above

Answer: d

8. It is thought that an asteroid may have struck the earth millions of years ago, leading to the extinction of the dinosaurs and many other species of plants and animals. Which of the following is NOT one of the aftereffects that is thought to have occurred after the asteroid struck the earth?

a. Fires were ignited which burned a lot of the world's forests.

b. The temperature on the earth became extremely hot for a period of several months.

c. An explosion equal to 10,000 times the world's current nuclear arsenal occurred.

d. All sunlight was blocked and the earth remained totally dark for a period of several months.

Answer: b

9. The respiratory structures used by amphibians include

a. gills.

b. lungs.

c. skin.

d. two of the above

e. all of the above

Answer: e

Requires understanding of respiration in amphibians.

10. Birds get rid of their nitrogenous waste (ammonia) by forming a substance known as

a. urine.

b. uric acid.

c. urea.

d. lactic acid.

e. ammonia.

Answer: b

In multiple-choice questions, be careful to distinguish among apparently related items, such as a, b, c, and d.

11. Some adult lampreys feed by

a. rasping holes in live fish.

b. capturing and eating snails and small fish.
c. filter feeding.
d. capturing and eating surfers.

Answer: a

12. The nitrogenous bases present in DNA are

a. adenine, guanine, cytosine, and thymine.
b. adenine, guanine, cytosine, thymine, and uracil.
c. adenine, guanine, cytosine, and uracil.
d. adenine, guanine, uracil, and thymine.

Answer: a

Memory is tested here. Multiple-choice questions that call for series, lists, or sequences must be carefully approached. Mistakes due to misreading are all too possible.

13. One characteristic of many freshwater fish that lived around 380 million years ago was

a. a simple lung.
b. a swim bladder.
c. they laid shelled eggs.
d. they were warm-blooded.

Answer: a

Calls for an understanding of vertebrate evolution.

14. The group Chordata is most closely related evolutionarily to the group

a. Arthropoda.
b. Echinodermata.
c. Mollusca.
d. Annelida.

Answer: b

Tests understanding of why systematists construct phylogenies in the way that they do.

15. You catch a cold. You get better, but your spouse catches your cold. Can you catch the cold back from your spouse?

a. yes

b. no

Answer: b

Memory cells of the immune system provide you with resistance to infective microorganisms to which you've been previously exposed. A major immunological process.

16. A gene is

a. composed of a chain of amino acids.

b. one of the four types of nucleotides found in DNA.

c. a portion of DNA in the chromosome that codes for a particular trait.

d. two of the above

e. all of the above (a, b, and c)

Answer: c

Requires understanding of a fundamental genetic concept.

17. The oldest fossils of *Homo sapiens* are thought to be about ______ years old.

a. 50,000

b. 300,000

c. 10 million

d. 100 million

e. 200 million

Answer: b

Requires grasp of evolutionary scale.

18. The "age of reptiles," when reptiles were the most common vertebrates on land, was

a. 550-400 million years ago.

c. 350-280 million years ago.

b. 400-350 million years ago.

d. 280-65 million years ago.

Answer: d

Calls for grasp of evolutionary scale.

19. Every time a human cell divides, the new cell that is produced has, on average, ______ change(s) in its DNA.

a. 1
b. 2
c. 3
d. 4
e. 5

Answer: a

Requires understanding of a fundamental concept in human reproduction and genetics.

20. The organs that are located in the roof of a snake's mouth used to detect scent are called

a. opercula.
b. Jacobson's organs.
c. gill slits.
d. pit organs.
e. gas glands.

Answer: b

21. Eastern and western meadowlarks are thought to have become separate species after

a. the Rocky Mountains separated them into eastern and western populations (groups).
b. a glacier separated them into eastern and western populations (groups).
c. the Grand Canyon separated them into two separate populations (groups).
d. the original species mated with a second species and produced a totally new species.

Answer: b

Requires understanding of evolutionary processes and how these are related to geological processes. Also calls for an appreciation of the scale of time involved in geologic and evolutionary change.

22. Which of the following is not a characteristic of *all* members of the Phylum Chordata?

a. Deuterostome development
b. Eucoelomate

c. Vertebrae made of bone or cartilage present during some stage of development
d. Post-anal tail present during some stage of development
e. All of the above are characteristics of all members of the Phylum Chordata.

Answer: c

Multiple-choice questions that include both a negative ("is *not*") and an absolute positive ("all") can be very tricky to answer correctly. Be certain you understand what the question is asking for before you respond.

23. It is thought that bacteria have become resistant to antibiotics because

a. antibiotics cause mutations to develop in the bacteria that make the bacteria resistant to the antibiotics.
b. antibiotics kill all bacteria except those that are slightly resistant to the antibiotic.
c. one species of bacteria will begin to mate with other species of bacteria if it is exposed to antibiotics, and give rise to new species of bacteria.
d. antibiotics cause bacteria to reproduce at a much faster rate than normal.

Answer: b

The student should understand that evolution acts on random mutations and differential reproduction, and that mutations do not occur in response to some need to adapt to a new environmental stress.

24. On the Galapagos Islands, iguanas (a type of lizard) feed mainly on

a. insects.
b. small vertebrates.
c. leaves.
d. seaweed.
e. fruit.

Answer: d

25. The species that causes the type of malaria present in Central Africa that commonly causes death in people is *Plasmodium*

a. *falciparum.*
b. *vivax.*
c. *malariae.*
d. *dubium.*

e. *singularis.*

Answer: a

26. Archaeopteryx had some reptilelike characteristics, such as

a. teeth.
b. solid bones.
c. a long tail containing vertebrae.
d. all of the above
e. none of the above

Answer: d

27. Which one of the following things is *not* thought to be caused by natural selection?

a. an increase in the number of individuals that have favorable traits
b. an increase in the rate at which DNA mutations occur in individuals
c. the origination of new species
d. change within species
e. a decrease in the number of individuals which have less favorable traits

Answer: b

Students often mistakenly think that an organism can somehow "will" evolution to occur.

28. Having three copies of the ______ chromosome results in Down syndrome in humans.

a. 3rd
b. 7th
c. 17th
d. 21st
e. 23rd

Answer: d

A test of recall that focuses on human genetics.

29. The genetic material of a virus may be either DNA or RNA.

a. true
b. false

Answer: a

30. One of the species of animal that Darwin saw on the Galapagos Islands that made him begin to think about natural selection was

a. a white moth with black speckles that blended in with the tree trunks on which it rested.

b. a species of bird (the cormorant), which had short stubby wings and could not fly.

c. a giant species of lizard with a very long neck (the Gila monster).

d. a species of bird thought to be extinct (the Dodo).

Answer: b

Tests recall of factual material in context.

31. In people infected with HIV, after they have been infected for a long time, the amount of HIV in their bodies increases and

a. the number of helper T red blood cells in their bodies increases.

b. the number of helper T red blood cells in their bodies decreases.

c. the number of helper T white blood cells in their bodies increases.

d. the number of helper T white blood cells in their bodies decreases.

Answer: d

Requires knowledge of an immune deficiency process.

32. Fossils of the earliest mammals show that they were the size of a

a. rat.

b. cat.

c. medium-sized dog.

d. cow.

e. *Tyranosaurus rex.*

Answer: a

33. Which of the following is characteristic of the Chondrichthyes?

a. Jaws can open 180 degrees.

b. They do not have scales.

c. Skeleton is composed mostly of bone.

d. all of the above

e. none of the above

Answer: a

34. The type of vaccine that we discussed in class that is given to humans to protect them from becoming infected with a virus that causes a particular disease may contain

a. blood cells infected with the virus.
b. a virus of the same type that has been modified so that it can't cause the disease.
c. antibiotics that kill the virus.
d. antibodies that kill the virus.

Answer: b

A correct response requires understanding of how vaccines work against viral (versus bacterial) infection.

35. What is the exact name of the external flap that covers over the gills on a bony fish?

a. operculum
b. placoid
c. gas gland
d. gill slit
e. pit organ

Answer: a

36. Sickle cell anemia is caused by

a. genes that code for altered hemoglobin that can cause red blood cells to sickle.
b. defective genes that cannot prevent malaria parasites from destroying red blood cells.
c. genes that code for sickle-shaped red blood cells.
d. genes that cause acid levels in the bloodstream to become abnormally high.

Answer: a

All of the alternatives sound plausible. A correct response requires an accurate understanding of the disease process in question.

37. Charles Darwin came upon his ideas about natural selection after finishing a trip

a. on a ship that went around the world.
b. on a ship that went only to South America and its associated islands.
c. on a ship that went only to Africa and its associated islands.
d. over a river and through some woods to visit his grandmother.

Answer: a

38. What is the name of the protein shell that encloses a virus's genetic material?

a. cell wall

b. capsule

c. capsid

d. viral envelope

e. none of the above

Answer: c

Tests recall of factual material, as well as a fundamental difference between viruses and cells (answer *a* is incorrect).

39. The scales found on sharks are known as

a. placoid scales.

b. ganoid scales.

c. ptenoid scales.

d. None of the above: sharks don't have scales.

Answer: a

Tests mastery of factual material. Many students respond with *d*.

40. Which of the following is a disease *not* caused by a virus?

a. flu

b. rabies

c. malaria

d. herpes

e. common cold

Answer: c

Tests understanding of bacterial versus viral diseases.

41. A codon

a. is a length of DNA that makes up a single chromosome.

b. is a length of DNA that contains the code to make one protein.

c. is a series of nucleotides in DNA that can carry the code for a particular amino acid.

d. is a type of RNA that can copy the genetic code from DNA.

Answer: c

Requires a basic grasp of genetic function at the molecular level.

42. Viruses cannot reproduce unless they infect a cell.

a. true

b. false

Answer: a

Requires understanding of a fundamental property of viruses.

43. In DNA, it takes ______ nucleotide(s) to code for a single amino acid.

a. 1

b. 2

c. 3

d. 4

e. 5

Answer: c

Requires understanding a fundamental feature of how DNA functions.

44. Most mammals are characterized by

a. having hair.

b. nourishing young with milk from mammary glands.

c. development of offspring within their mother's uterus.

d. all of the above

e. none of the above

Answer: d

Requires grasp of fundamental mammalian characteristics.

45. Gene mutations

a. may be harmful to the organism that has the mutation.

b. may be helpful to the organism that has the mutation.

c. may have no effect on the organism that has the mutation.

d. all of the above

e. none of the above

Answer: d

Know the material. Students often are predisposed to believe that mutations are always harmful.

46. A person who lives in an area where malaria is common and where no sickle cell anemia drugs or antimalaria drugs are available would be most likely to live a long life if he or she had the genes

a. HH.

b. Hh.

c. hh.

d. two of the above

e. none of the above

Answer: b

An HH individual will produce all normal hemoglobin, but will not be resistant to malaria and will thus likely contract it. An hh individual will be resistant to malaria, but will have sickle cell anemia (SCA), because all of the hemoglobin that he or she produces will cause RBCs to sickle in an acidic environment. The Hh individual will not have SCA and will be resistant to malaria. Enough altered hemoglobin will be produced to cause RBCs to sickle as they become acidic internally as *Plasmodium* reproduces, but massive sickling of all RBCs will not occur, because of the production of normal hemoglobin. As *Plasmodium*-infected RBCs sickle, they will be phagositized by macrophages. Thus the bloodstream will be cleaned of *Plasmodium*, but massive sickling does not occur.

47. On average, about how long does it take before an adult infected with HIV develops AIDS (in the absence of any drug treatment for the disease)?

a. $1^1/_2$ years

b. 3 years

c. 5-10 years

d. 20 years

Answer: c

48. Gene mutations only occur in people (and other organisms) exposed to excessive radiation (e.g., 100s of X-rays, radioactive waste, etc.).

a. true

b. false

Answer: b

Students are predisposed to believe that mutations occur only when organisms are exposed to certain harmful substances.

49. Which one of the following statements is *not* one of the parts of Darwin's theory of natural selection?

a. Among a given species, some individuals may have a better chance for survival than others.

b. All organisms show variation in traits.

c. In nature, the majority of organisms produce offspring only if the offspring are likely to survive.

d. New species originate through natural selection.

Answer: c

Students tend to apply to the natural world the human perspective: only have children if you are likely to be able to support them.

50. An animal dies with 1 percent of the carbon in its body in the form of carbon-14; 11,200 years later, you would expect to find ______ percent of the carbon in the fossil to be carbon-14.

a. $^1/_8$

b. $^1/_4$

c. $^1/_2$

d. 2

e. 4

Answer: b

The correct response combines mastery of factual material with an understanding of how radioactive decay is used to date fossil material. The half-life of carbon-14 is 5,100 years. If something died 11,200 years ago with 1 percent of the carbon in its body in the form of carbon-14, then 5,100 years ago, this would be halved, or $^1/_2$ percent. As another 5,100 years passes, bringing us to the present, this $^1/_2$ percent would, in turn, be halved, leaving us with $^1/_4$ percent carbon-14.

CHAPTER 22

NEW YORK UNIVERSITY

V23.0011, V23.0012: PRINCIPLES OF BIOLOGY

Richard Borowsky, Associate Professor

THIS IS A TWO-SEMESTER COURSE. SIX EXAMS ARE GIVEN DURING THESE TWO SEMESters, two midterms and one final per semester. All are multiple-choice exams. A sample final exam is presented here. Together, the exams count for 70 percent of the course grade, 20 percent for each midterm and 30 percent for the final.

Our goal is for students to learn the fundamentals of biology during these two semesters, covering all areas. Students preparing for MCAT exams should get all the biology material relevant to the standardized exam.

We hope to impart the knowledge base and analytical skills to appreciate the importance of biological science as part of our general culture as well as its practical significance (medicine, agriculture, conservation, etc.). After taking this course, students should also have the ability to evaluate new findings critically and scientifically and to be fortified against the growing trends of superstition and the negative view of science.

The idea is that the course objectives should be met through our lectures, not our exams. Primarily, the exams are designed to test how closely students listened to the lectures and how much thought they gave the material afterward. That is, the course examinations are designed more as a test of knowledge and assimilation of lecture topics than as a teaching tool. We expect the students to remember facts, to master analytical techniques, and to be able to synthesize. The exam questions are heavily biased toward fact retention and association. This is easiest for most students and separates those who have studied from those who have not. These questions establish the base grade for the class. The analytical and integrative questions separate the good students from the mediocre.

The best way to succeed on the exams and in the course is to put pen to paper. Take good notes in class and redraw—from memory—all relevant figures found in the textbook. If you own a tape recorder, do not take it to class! Give the tape recorder to your worst enemy.

Study lecture notes and read the book within 24 hours of attending the lecture. If you start studying for the exam two weeks in advance, you are probably already four weeks too late.

FINAL EXAM

1. Which structure is characteristic of the cells or tissues of both plants and animals?

a. gap junctions
b. chloroplasts
c. nuclei
d. specialized tissues for coordinated movement
e. desmosomes

Answer: c

2. The archenteron is

a. the basis of the gut cavity.
b. the primitive coelom.
c. schizocoelous in development.
d. found in all eukaryotes at some stage in development.
e. a high church official having diverse duties, both secular and spiritual.

Answer: a

3. Invertebrate species are believed to make up close to what percentage of all animal species?

a. 10%
b. 30%
c. 50%
d. 70%
e. 90%

Answer: e

4. An animal phylum is defined on the basis of

a. the uniqueness of its DNA content.
b. the uniqueness of its body plan.
c. its type of body plan symmetry.
d. its fundamental habitat (aquatic, terrestrial, etc.).
e. its physiological specializations.

Answer: b

5. The animal phyla that exist today

a. came into being over a 300-million-year period starting 700 million years ago (MYA).
b. came into existence over a relatively short period in the Cambrian, about 550 MYA.
c. left extensive fossil evidence starting 700 MYA.
d. already existed by the start of the Cambrian.
e. did not leave extensive fossil evidence until the end of the Cambrian.

Answer: d

6. Biologists interested in the evolutionary relationships of organisms try to identify traits called "synapomorphies." A synapomorphy is defined as

a. a derived character that is unique to a taxon.
b. an ancestral character that is unique to a taxon.
c. a character that is shared by all taxa.
d. an ancestral character that is shared by all taxa.
e. a derived character that is shared by some taxa.

Answer: e

7. Which characteristic of many different animal groups is associated with bilateral symmetry?

a. cephalization
b. sessile life style
c. autotrophy
d. radial body plan
e. contractile fibers (muscle or musclelike)

Answer: a

8. Which of the following invertebrate classes or phyla lacks a gut cavity in the adult form?

a. Rotifera

b. Nematoda
c. Polyplacophoran mollusc
d. Cestode platyhelminth
e. Polychaete annelid

Answer: d

9. Deuterostomes have

a. two mouths.
b. no coelomic cavity.
c. radical development.
d. enterocoelous development.
e. schizocoelous development.

Answer: d

10. Which insect order has the most species?

a. Diptera—the flies
b. Coleoptera—the beetles
c. Hymenoptera—the wasps
d. Hemiptera—the bugs
e. Lepidoptera—the butterflies

Answer: b

11. Which of the following is a key evolutionary innovation of the Insecta?

a. the keratinized cuticle
b. forelimbs developed as wings
c. segmented body form
d. winged flight
e. the so-called "six-leg body plan"

Answer: d

12. The onycophoran "walking worm" is important because

a. while small, the combined biomass of all individuals of the species far exceeds that of earthworms and they are a major force in the conditioning of soils, especially in England.
b. they lack segmentation yet have a molluscan mantle.
c. they have segmentation but lack legs and are considered ancestral to both annelids and arthropods.
d. their body plan suggests that of an intermediate between those of other major phyla.

e. while cephalized, they lack brains.

Answer: d

13. The nematode *Caenorhabditis elegans* is an important model organism for the study of developmental biology because

a. it exhibits extreme developmental plasticity.
b. although small, individuals have millions of cells to study.
c. its opaque outer covering greatly limits damaging UV-induced mutations.
d. it has a fixed number of cells with normally inflexible developmental fates.
e. mutations are known, which turn its balancing organs (halteres) into an extra pair of wings.

Answer: d

14. The cercaria larva

a. is one of two larval forms of trematode parasites (Phylum Platyhelminthes), and infects vertebrate hosts by burrowing into the host's skin.
b. is a sexually reproducing form of cestode.
c. is the dispersal stage of most, but not all, marine invertebrates.
d. is considered a delicacy in Pacific Island cultures.
e. is a life stage of Obelia and other marine hydrozoans.

Answer: a

15. The medusa is

a. the dominant stage among the Scyphozoa (jellyfish).
b. the dominant stage among the anthozoan corals.
c. the dominant stage among the anthozoan anemones.
d. the dominant stage among all of the hydrozoans.
e. a high church official having principally secular duties.

Answer: a

16. A Parazoa is

a. two animals.
b. a subkingdom that contains the sponges.
c. an animal with a radial body plan.
d. not even an animal.
e. a geometrically defined curve.

Answer: b

17. A pseudocoelom is

a. found in all bilateria.
b. found in all coelomates.
c. not found in annelid worms.
d. a digestive cavity with only one opening.
e. a digestive cavity with two openings, but only one is lined with mesoderm.

Answer: c

18. A blastula is a stage of development

a. common to all eukaryotic organisms.
b. typical of animal development only.
c. consisting of a solid ball of cells.
d. consisting of a hollow ball of cells with an invagination, similar in structure to a balloon with a finger pressed into it.
e. not worth your thought.

Answer: b

19. It is hypothesized that animals evolved from colonial

a. bacteria.
b. fungi.
c. plants.
d. flagellates.
e. viruses.

Answer: d

20. Our closest living invertebrate relatives are

a. starfish.
b. nematodes.
c. chordates.
d. annelids.
e. reptiles.

Answer: c

21. Which of the following is not a character uniting all chordates?

a. notochord
b. spinal column
c. dorsal, hollow nerve tube
d. pharyngeal slits
e. segmented, muscular, post-anal tail

Answer: b

22. Vertebrates generally have all of the following characteristics, except
a. cephalization.
b. spinal columns.
c. a chitinous endoskeleton.
d. a closed circulatory system.
e. an appendicular skeleton.
Answer: c

23. Paedogenesis is
a. attainment of sexual maturity in the larval stage.
b. a stage of development in the tunicata.
c. loss of a character in the adult stage of chordates.
d. a stage of development in echinoderms.
e. a serious crime, not much discussed openly.
Answer: a

24. Which of the following groupings of animal species has the least taxonomic validity?
a. birds
b. mammals
c. amniotes
d. fishes
e. tetrapods
Answer: d

25. The basic adaptation of birds is winged flight. Which one of the following structural changes in birds is not closely related to flight?
a. good hearing
b. good eyesight
c. good muscular coordination
d. hollow bones
e. loss of teeth and jaw bones
Answer: a

26. All mammals have all of the following traits, except
a. amnion.
b. chorion.
c. placenta.
d. hair.
e. milk.
Answer: c

27. Our most distant ape relatives are

a. gorillas.
b. orangutans.
c. gibbons.
d. chimpanzees.
e. howler monkeys.

Answer: c

28. Which one of the following is not a general property of living organisms?

a. Their organization has a cellular basis.
b. They are hierarchically organized.
c. They reproduce their own kind.
d. They have two sets of chromosomes (homologues) in each somatic cell.
e. They grow and develop.

Answer: d

29. One "emergent property" at the level of organism is

a. specific interactions among molecules.
b. cell division.
c. reproduction.
d. evolution.
e. ecology.

Answer: c

30. Reductionism is an approach to biology that yields very satisfying answers to questions we have about life. Holism is

a. a way of seeing what the important questions are in the first place.
b. another name for vitalism.
c. simply another name for reductionism.
d. an approach to science, in general, that is nearly universally accepted.
e. a genocentric view of the organism in its surroundings.

Answer: a

31. Which one of the following categories of taxonomic classification is the least inclusive?

a. class
b. family
c. genus

d. order
e. phylum

Answer: c

32. **Campbell uses the term "unity in diversity" to describe a seeming contradiction: living beings are incredibly diverse in form and function, yet all share certain characteristics, such as a nucleic acid information storage system. Unity in diversity is best explained, as follows:**
a. There is only one way to design a living system.
b. Although life originated many times, each origination was on the same planet.
c. Life on Earth was seeded from a single source from another galaxy.
d. Life on Earth was seeded from a single source within our own galaxy.
e. All living things on Earth share a common ancestor.

Answer: e

33. **Constructing hypotheses and devising tests capable of proving them wrong is the core technique of**
a. science.
b. law.
c. sales.
d. politics.
e. business.

Answer: a

34. **Cave organisms generally possess a set of features known as "troglomorphic characteristics." Which of the following is not one of these characteristics?**
a. loss of vision
b. atrophy of the brain's sensory areas
c. longer legs and antennae
d. lower or more efficient metabolism
e. an increase in acuity of most senses

Answer: b

My specialty is cave fishes, so students *hear* about this!

35. **Regressive evolution is**
a. only found in cave animals.
b. only truly found in parasites.

c. an increase in sensory capacity to compensate for a loss of vision.
d. a switch to nonvisual signaling mechanisms when there is a loss of pigment or vision.
e. the evolutionary loss of or decrease in a structure.

Answer: e

36. Evolution is the emergent property of

a. macromolecules.
b. cells.
c. organisms.
d. populations.
e. ecosystems.

Answer: d

37. Cell reproduction in bacteria is accomplished by

a. meiosis.
b. mitosis.
c. endocytosis.
d. exfoliosis.
e. binary fission.

Answer: e

38. To a cytogeneticist, a chromosome is an entity containing chromatin that

a. has a sister chromatid.
b. contains two sister chromatids.
c. only has a single chromatid.
d. has a centromere.
e. has a homologue.

Answer: d

39. The stage of mitosis in which chromosomes separate is

a. interphase.
b. anaphase.
c. prophase.
d. telophase.
e. metaphase.

Answer: b

40. A cell has eight chromosomes at the start of interphase. How many chromatids does it have at the start of metaphase I?

a. 2
b. 4
c. 8
d. 16
e. 32

Answer: d

41. The phase of the mitotic cycle in which DNA is replicated is
a. G_1 of interphase.
b. G_2 of interphase.
c. S of interphase.
d. anaphase.
e. metaphase.

Answer: c

42. Kinetochore fibers are
a. structures that only exist up to the metaphase stages of mitosis.
b. microfilament structures that are present in dividing cells during mitosis and meiosis.
c. microtubules that attach chromosomes at their centromeres to centrosome regions during cell division.
d. found within duplicated chromosomes and run from the kinetochore of one sister chromatid to that of the other. They bind the two sisters together and are severed prior to the appropriate anaphase.
e. longer than nonkinetochore fibers.

Answer: c

43. The essential observation made by Mendel in his experiments on inheritance in pea plants was that
a. parental characters that were not exhibited in the F_1 generation reappeared in the F_2.
b. genetic information was encoded in DNA sequence.
c. a 3:1 phenotypic ratio (dominant:recessive) was exhibited by the F_2 generation.
d. a 3:1 phenotypic ratio (recessive:dominant) was exhibited by the F_2 generation.
e. factors for the development of traits, rather than traits themselves, were inherited.

Answer: a

44. The pea plant has a haploid number of 7. What is the probability that any given pollen granule produced by a pea plant will contain all seven centromeres derived from its male parent's chromosomes?

a. 1
b. 1/7
c. $1/(2^7)$
d. $1/(7^2)$
e. smaller than $1/(2^7)$

Answer: c

45. The pea plant still has a haploid number of 7. What is the probability that any given pollen granule produced by a pea plant will contain all seven chromosomes derived from its male parent's chromosomes?

a. 1
b. 1/7
c. $1/(2^7)$
d. $1/(7^2)$
e. smaller than $1/(2^7)$

Answer: e

This is because of recombination.

46. Homologues separate during

a. anaphase I
b. metaphase I
c. anaphase II
d. prophase II
e. mitosis

Answer: a

47. The "response to selection" observed by animal and plant breeders in populations or herds artificially selected for some trait is

a. the change in life span with increasing stringency of the environment.
b. fastest in a genetically uniform population.
c. only possible in a genetically uniform population, although it may be slow or fast.
d. a shift in the average phenotype from one generation to the next.
e. sustainable for an infinite number of generations.

Answer: d

48. Mendel's success in elucidating the mechanism of inheritance in most sexually reproducing organisms was attributable to all of the following, save one:

a. He chose a model organism that is easy to grow.

b. Pea plants are either male or female. Therefore, self-fertilization does not occur in peas.

c. He studied traits that were discrete and easily classified.

d. He used quantitative methods.

e. He worked with inbred strains that bred true.

Answer: b

49. In peas, tall plant size is dominant to short, purple flower color is dominant to white, and yellow seed color is dominant to green. All three of these traits are independently inherited. For simplicity's sake, the alleles for the six characteristics are designated T (tall), S (short), P (purple), W (white), Y (yellow), and G (green). A plant of genotype TS PW YG is mated to a plant that is short and has purple flowers and green seeds. What proportion of their offspring will be tall and have purple flowers and yellow seeds?

a. $3/8$

b. $1/2$ or $1/4$

c. $3/8$ or $1/2$

d. $3/16$ or $1/4$

e. $1/16$

Answer: d

50. A pea plant of unknown genotype is crossed with another pea plant that is tall and has purple flowers and yellow seeds. With respect to determining the genotype of the unknown for height, flower color, and seed color, this cross is

a. a testcross.

b. a backcross.

c. worthless.

d. the needed preliminary for making an F_1.

e. revealing for all three characteristics.

Answer: c

This is not a backcross; we cannot resolve the genotype in this case.

51. In certain species of freshwater fishes (the platyfishes), pigmentation variation is controlled by alleles at a single locus. Individuals with different pigment patterns also have different metabolic rates. The correlation of metabolic rate with pigmentation is an example of

a. multiple alleles.
b. codominance.
c. incomplete dominance.
d. pleiotropy.
e. polygenic inheritance.

Answer: d

52. Variation in human height is believed to be under the control of alleles at several different loci. Each of these loci is hypothesized to have two alleles, one for taller and the other for shorter. The overall phenotype is determined by the total number of "tall" alleles summed over all the height loci. Systems like this are usually designated

a. multiple allelic.
b. codominant.
c. incompletely dominant.
d. pleiotropic.
e. polygenic.

Answer: e

53. Which one of the following distinguishes prokaryotic cells from eucaryotic cells? Prokaryotes

a. have cell walls but not cell membranes.
b. have mitochondria.
c. lack membrane-enclosed nuclei.
d. have nuclei with no DNA.
e. lack ribosomes.

Answer: c

54. Which of the following levels in the hierarchy of biological organization includes all of the other levels in the list?

a. cells
b. biological molecules
c. atoms
d. tissues
e. organelles

Answer: d

55. With each step upward in the hierarchy of biological order, novel properties emerge that were not present at the simpler levels of organization. These emergent

properties result from

a. vital forces that arise at each level.

b. the arrangement and interactions between components.

c. the physical and chemical phenomena that operate only in living things.

d. simple summation of the individual behavior of the component parts.

e. the emergence of supernatural forces.

Answer: b

Watch out for throwaway responses, such as e, above. Be sure to eliminate them first from among your multiple choices!

56. Two species belonging to the same class must also belong to the same

a. species.

b. order.

c. genus.

d. family.

e. phylum.

Answer: e

57. According to the fluid-mosaic model of cell membranes, which of the following is a true statement about membrane phospholipids?

a. They frequently flip-flop from one side of the membrane to the other.

b. They move laterally along the plane of the membrane.

c. They are free to depart from the membrane and dissolve in the surrounding solution.

d. They have hydrophilic tails in the interior of the membrane.

e. They occur in an uninterrupted bilayer, with membrane proteins restricted to the surface of the membrane.

Answer: b

58. The movement of uncharged molecules from a low concentration to a higher concentration is described by which of the following?

a. exocytosis

b. osmosis

c. facilitated diffusion

d. active transport

e. diffusion

Answer: d

59. What is the voltage across cell membranes called?

a. osmotic potential
b. chemical gradient
c. water potential
d. membrane potential
e. electrochemical gradient

Answer: d

If you have to guess among multiple choices, look for the obvious clues. Here the word *membranes* is in the question and only in one response, *d,* making it a prime candidate from among the other choices.

60. The movement of potassium into or out of the cell requires

a. glucose for binding and releasing ions.
b. plant hormones embedded in the cell membrane.
c. high cellular concentrations of potassium.
d. ATP as an energy source.
e. low cellular concentrations of sodium.

Answer: d

61. An organism with a cell wall would be unable to do which process?

a. osmosis
b. phagocytosis
c. active transport
d. diffusion
e. exocytosis

Answer: b

62. The kinds of molecules that pass through a cell membrane most easily are

a. small and hydrophobic.
b. ions.
c. monosaccharides such as glucose.
d. large and hydrophobic.
e. large polar molecules.

Answer: a

63. What membrane-surface molecules help cells recognize each other?

a. integral proteins

b. peripheral proteins
c. carbohydrates
d. cholesterol
e. phospholipids

Answer: c

64. Which of the following organisms does not reproduce cells by mitosis?

a. bacterium
b. mushroom
c. cockroach
d. cow
e. banana

Answer: a

65. If there are 12 chromosomes in an animal cell in the G_2 stage of the cell cycle, what is the diploid number of chromosomes for this organism?

a. 48
b. 6
c. 12
d. 24
e. 36

Answer: c

66. Measurements of the amount of DNA per nucleus were taken on a large number of cells from a growing fungus. The measured DNA levels ranged between 3 to 6 picograms (pg) per nucleus. One nucleus had 5 pg of DNA. What stage of the cell cycle was this nucleus in?

a. G_0
b. G_1
c. M
d. G_2
e. S

Answer: e

67. Centromeres uncouple, sister chromatids are separated, and each new chromosome moves to opposite poles of the cell during

a. telophase.
b. anaphase.

c. metaphase.
d. prophase.
e. interphase.

Answer: b

68. During which phase of mitosis are two-chromatid chromosomes visible?
a. from metaphase through telophase
b. from G_2 of interphase through metaphase
c. from the end of interphase until anaphase
d. from G_1 of interphase through metaphase
e. from anaphase through telophase

Answer: c

69. If there are 10 chromosomes in a cell, how many centromeres are there?
a. 5
b. 10
c. 20
d. either 5 or 10
e. either 10 or 20

Answer: b

70. White eyes in *Drosophila melanogaster*
a. is inherited as a sex-linked recessive.
b. is not under genetic control.
c. led Mendel to the discovery of pedigree analysis.
d. is a "wild-type."
e. is inherited as a sex-linked dominant.

Answer: a

71. Alleles of linked genes
a. are functionally related to one another.
b. are only found in eukaryotes.
c. were discovered by Mendel in his garden.
d. tend to cosegregate in a backcross of a dihybrid.
e. are never found on the same chromosome.

Answer: d

72. The basis of genetic recombination in higher organisms is most often
a. an exchange of information between DNA molecules unaccompanied by any

physical exchange of chromatin.
b. a physical recombination of material between nonhomologous chromosomes.
c. a physical recombination of chromosomal material between homologous chromatids.
d. a physical recombination of chromosomal material between sister chromatids.
e. the exchange of histones, but not DNA, between any types of chromosomes.

Answer: c

73. Four genetic markers on chromosome III of *Drosophila melanogaster* are studied in a series of dihybrid crosses. It is determined that the distances between pairs of loci are:

A to B	5 cm
A to C	11 cm
B to C	7 cm
B to D	6 cm

Which one of the following is true?
a. The order of loci could be A_B_C_D.
b. The order of loci could be A_B_D_C.
c. The order of loci could be B_A_D_C.
d. The order of loci is unambiguously determined by the data as D_A_B_C.
e. The order of loci is unambiguously determined by the data but cannot be determined by the test taker.

Answer: b

Avoid *e*. No points for honesty!

74. Most calico cats are female. Occasionally a male calico is observed. How many Barr bodies would you expect to see in cells from male calico cats?
a. 0
b. 1
c. 2
d. 0 or 1
e. 1 or 2

Answer: b

75. The frequency of which of the following aneuploid disorders increases with increasing age of the mother?
a. Turner syndrome
b. Klinefelter syndrome

c. Trisomy X (XXX or metafemale)
d. XYY (extra Y male)
e. Down syndrome

Answer: e

76. The leading cause of spontaneous abortion is

a. aneuploidy of the fetus.
b. triploidy of the fetus.
c. haploidy of the fetus.
d. women conceiving too late in life.
e. environmental pollution.

Answer: a

In the context of questions about genetics, it is unlikely that *d* or *e* could be correct. Eliminate them.

77. A man who is afflicted with hemophilia and a woman who is a carrier but not a hemophiliac have children. Which of the following is a correct statement about their children?

a. All the males and half of the females are carriers.
b. All the males are hemophiliac.
c. All the males carry the gene for the disease, but only half of them are hemophiliac.
d. All the females carry the gene for the disease, but only half of them are hemophiliac.
e. All the offspring carry the gene for the disease, but only half of them are hemophiliac.

Answer: d

78. In the ABO blood group system, the allele for type O has a population frequency of 0.65. A woman (type B) has a baby (type O) and accuses a man (type A) of being the father. Another man, the only other one who might be the father—both had equal access to the mother—has enrolled in a Federal Witness Relocation Program. The paternity testing laboratory is having some difficulty finding him and arranging for a blood test. What can you say about the first man, based only on the blood test results you have and the population frequency of the allele for O?

a. It shows he can't be the father.
b. It shows he must be the father.
c. It shows he is more likely to be the father than is the second man.

d. It shows he is less likely to be the father than is the second man.
e. It shows he is as likely to be the father than is the second man.

Answer: d

The likelihood ratio is $\frac{P_1 < .5}{P_2 = .65} < \frac{1}{1}$

where P_1 is the probability of getting type O out of the first man, and P_2 is the probability of getting type O out of the second man.

79. In the list given, the first point at which you could distinguish between a mitosis and a meiosis would be

a. telophase.
b. anaphase.
c. metaphase.
d. prophase.
e. prophase II.

Answer: d

80. The map of the bacterial chromosome that is measured in time units (minutes) is based on data from

a. mutation studies.
b. specialized transduction studies.
c. generalized transduction studies.
d. transformation studies.
e. bacterial mating studies.

Answer: e

81. Which one of the following statements about viruses is true?

a. Viruses are the most primitive life-forms.
b. Viruses are certainly primitive life-forms, but not the most primitive.
c. All viruses have DNA cores, although the DNA can be single-stranded or double-stranded.
d. Viruses have a nucleic acid-based genome enclosed in a protein coat.
e. Viruses all have simple prokaryotic gene structure.

Answer: d

82. The HIV virus

a. has a DNA core.
b. inserts an RNA copy of its chromosome into mRNA transcripts of its host cells and thereby ensures that its message is translated.
c. makes a DNA copy of its chromosome which is inserted into the host genome similar to a provirus.
d. makes an RNA copy of its chromosome which is inserted into the host genome similar to an antivirus.
e. is easily eradicated by antibiotic treatment.

Answer: c

83. A viral plaque is
a. best fought with viral toothpaste.
b. a patch of dead bacteria on a lawn.
c. the stiff coating on the outside of a virus that protects its DNA core and allows it to go on living in a hostile and threatening world.
d. a patch of dead viruses on a green.
e. only caused by "T-even" bacteriophage.

Answer: b

84. A lytic infection in the reproduction of phage lambda
a. involves the insertion of the circularized phage into the host chromosome.
b. involves the insertion of a copy of the circularized phage into the host chromosome.
c. involves the insertion of the linearized phage into the host chromosome.
d. involves the insertion of a copy of the linearized phage into the host chromosome.
e. produces many daughter phage particles from a single host cell.

Answer: e

85. Episomes have a functional similarity to which one of the following?
a. retroviruses
b. lysogenic phage
c. the bacterial chromosome
d. sex chromosomes
e. autosomes

Answer: b

86. A tryptophan-requiring bacterium is maintained in pure culture in a tryptophan-supplemented nutrient broth. An aliquot is plated out on sterile minimal medium,

and a few colonies of the bacterium are observed to grow. What is the most likely explanation for their growth?

a. Someone sneezed on the plate.
b. mutation
c. recombination caused by mating with another strain
d. transduction
e. transformation

Answer: b

87. Which of the following processes is essential for bacterial transformation to occur?

a. mutation
b. specialized transduction
c. generalized transduction
d. recombination
e. mating between strains

Answer: d

88. Anabolic pathways in prokaryotes generally

a. are constitutive.
b. control the stepwise degradation of complex sugars.
c. are controlled by inducible genes organized into operons.
d. are controlled by repressible cotranscribed genes.
e. are uncontrolled and wasteful of resources.

Answer: d

89. Transcription of the genes for the enzymes needed to utilize lactose in *Escherichia coli* exemplifies

a. positive control.
b. negative control.
c. cotranscription.
d. induction.
e. all of the above

Answer: e

90. A mutation in the gene that codes for the repressor protein of the trp operon in *Escherichia coli* that has a single effect—it renders the protein incapable of binding to tryptophan—would cause which of the following bacterial phenotypes?

a. a bacterium that can't turn the trp operon on

b. a bacterium that can't turn the trp operon off
c. Tryptophan will always induce the trp operon.
d. Tryptophan will always repress the trp operon.
e. Tryptophan will always cancel the operon's constitutive nature.

Answer: b

91. The nucleosome structure of eukaryotic chromatin is due to all of the following histones, except

a. H1.
b. H2A.
c. H2B.
d. H3.
e. H4.

Answer: a

92. Gamma chain globins are part of the beta globin family and are a relatively late development in the evolution of tetrameric hemoglobin. They form hemoglobins that function efficiently in low-oxygen-tension environments. They are found in which animal group(s)?

a. lizards
b. birds
c. mammals
d. a and c
e. b and c

Answer: c

93. The opportunities for control of gene expression are greater in eukaryotes than in prokaryotes. Which one of the following does not contribute to this greater diversity of control points in eukaryotes?

a. variable RNA transcript splicing
b. mRNA transport to the cytoplasm
c. coupling of transcription and translation
d. highly variable rates of transcript degradation
e. transcript capping and tailing

Answer: c

94. Coordinate transcription of genomically dispersed, but functionally related genes in eukaryotes is aided by their sharing

a. unique promoter sequences.

b. associated unique enhancer sequences.
c. associated unique suppressor sequences.
d. polypeptide products.
e. unique introns.

Answer: b

95. Steroid receptors are proteins

a. that can bind to DNA.
b. that can bind to steroids.
c. that can bind to inhibitory proteins.
d. all of the above
e. none of the above

Answer: d

96. Cells that have stopped dividing and are differentiating are

a. in the S phase of the cell cycle.
b. in the G_0 phase of the cell cycle.
c. cancer cells.
d. in the M phase of the cell cycle.
e. in the G_1 phase of the cell cycle.

Answer: b

97. All of the following occur during mitosis, except

a. the coiling of chromosomes.
b. the pairing of homologous chromosomes.
c. the formation of a spindle.
d. the division of centromeres.
e. the degradation of the nuclear envelope.

Answer: b

98. Which of the following events occurs during prophase I of meiosis?

a. synapsis and crossing over
b. segregation of alleles of unlinked genes
c. duplication of chromatids
d. segregation of alleles of linked genes
e. reduction in chromosome number

Answer: a

99. How do cells at the completion of meiosis compare with the cell from which they were derived?

a. They have the same number of chromosomes and half the amount of DNA.
b. They have half the amount of cytoplasm and twice the amount of DNA.
c. They have half the number of chromosomes and half the amount of DNA.
d. They have twice the amount of cytoplasm and half the amount of DNA.
e. They have the same number of chromosomes and the same amount of DNA.

Answer: c

100. What is a human cell that contains 22 pairs of autosomes and two X chromosomes?

a. an unfertilized egg cell
b. a sperm cell
c. a male somatic cell
d. a female somatic cell
e. an aneuploid

Answer: d

101. Crossing over occurs during which phase of meiosis?

a. telophase I
b. prophase I
c. anaphase I
d. metaphase II
e. prophase II

Answer: c

102. What is a genetic cross between a homozygous recessive individual and one of an unknown genotype?

a. a dihybrid cross
b. a self-cross
c. a test-cross
d. an F_1 cross
e. a hybrid cross

Answer: c

103. Feather color in budgies is determined by two different genes that affect the pigmentation of the outer feather and its core. Y_B_ is green; yyB_ is blue; Y_bb is yellow; and yybb is white. (Y_ means YY or Yy, B_ means BB or Bb.) A green budgie is crossed to a blue budgie and four young are obtained. Which one of the

following is true?

a. All four offspring cannot be yellow.
b. All four offspring cannot be blue.
c. All four offspring cannot be green.
d. All four offspring cannot be white.
e. Each one of the young can be any color.

Answer: e

104. What was the most significant conclusion that Gregor Mendel drew from his research?

a. There is considerable genetic variation in garden peas.
b. An organism that is homozygous for many recessive traits is at a disadvantage.
c. Genes are composed of DNA.
d. Dominant genes occur more frequently than recessive ones.
e. Factors for the development of traits are inherited in discrete units, one from each parent.

Answer: e

105. Given the parents AABBCc × AabbCc, assume simple dominance (A over a, B over b, C over c) and independent assortment. What proportion of the progeny will be expected to phenotypically resemble the first parent?

a. 1/4
b. 3/4
c. 1
d. 3/8
e. 1/8

Answer: b

106. The fact that all seven of the garden pea traits studied by Mendel obeyed the principle of independent assortment means that the

a. seven pairs of alleles determining these traits are on the same pair of homologous chromosomes.
b. seven pairs of alleles determining these traits behave as if they are on different chromosomes.
c. formation of gametes in plants is by mitosis only.
d. haploid number of garden peas is 7.
e. diploid number of garden peas is 7.

Answer: b

107. A Barr body (sex chromatin) is normally found in the nucleus of which kind of human cell?

a. sperm cells only
b. unfertilized egg cells only
c. somatic cells of a male only
d. somatic cells of a female only
e. both male and female somatic cells

Answer: d

108. The particular position of a gene on a chromosome is known as a(n)

a. map distance.
b. locus.
c. chiasma.
d. allele.
e. tetrad.

Answer: b

109. The frequency of crossing over between any two linked genes is

a. the same as if they were not linked.
b. proportional to the distance between them.
c. difficult to predict.
d. more likely if they are recessive.
e. determined by their relative dominance.

Answer: b

110. A man who carries an X-linked allele will pass it on to

a. all of his sons.
b. all of his children.
c. all of his daughters.
d. half of his daughters.
e. half of his sons.

Answer: c

111. Thomas Hunt Morgan is best remembered for his discovery and explanation of

a. dominance.
b. the Lyon hypothesis.
c. sex linkage.

d. crossing over and recombination.
e. extranuclear inheritance.

Answer: c

Read the question carefully and take from it all it has to offer. A man named Morgan will not be responsible for "the Lyon hypothesis." You've eliminated one choice.

112. There is good evidence for linkage when

a. two genes occur together in the same gamete.
b. two characteristics are caused by a single gene.
c. two genes work together to control a specific trait.
d. a gene is associated with a specific phenotype.
e. two different loci do not segregate independently during meiosis.

Answer: e

CHAPTER 23

UNIVERSITY OF PITTSBURGH

BIO 150 AND 160: FOUNDATIONS OF BIOLOGY I AND II

Laurel Roberts, Lecturer

FOUR EXAMS ARE GIVEN IN THIS COURSE, THREE "MIDTERMS" AND A FINAL. ALL ARE multiple choice.

This two-semester sequence is a survey course mainly for prehealth, biology, and other science majors; therefore, my primary objective is to provide a good introduction to as many areas of biology as possible. The courses are taught in multiple sections (usually four) with approximately 250 students enrolled per section.

I expect students to master a body of information on cell structure, genetics (especially molecular), and evolution. Students should be able to organize material well enough to relate it to analogies and to make real-world applications. Students should be able to answer correctly compare-and-contrast questions.

The course grade is wholly dependent on exam performance, with the three midterms accounting for a total of 75 percent of the grade, and the final for 25 percent.

I advise using "the 24H Turnaround" study tactic. Within twenty-four hours of each lecture, the student should review the lecture material, using the text as a reference. I usually recommend that they also recopy their notes for clarity. Weekly, take turns presenting synopses of the lecture topics to a study partner or study group. Also, make a review sheet each week from lecture topics. Write—but don't answer—sample questions. It is also helpful to go over exam review sheets, old exams, and sample questions.

MIDTERM EXAM

The multiple-choice questions in this exam fall mainly into three categories, those requiring factual knowledge, those requiring conceptual understanding, and those calling for a real-world application of principles learned. For many of the questions that follow, I have noted the type.

I try to include a number of analogy questions that relate biological processes to nonbiological phenomena. I do this to determine whether students have reached the synthesis stage, where they see the given concept as a whole rather than as random bits of information. I also tend to ask many factual questions, although they are usually less intricate than the lecture material would suggest.

1. Which of the following bonds is disrupted by hair perming or wool washing?
   a. ionic
   b. platonic
   c. covalent
   d. hydrogen
   e. hydrophobic

Answer: d

An opportunity to apply to a real-world situation something learned in class.

2. "The energy actually available within a system" is a description of which of the following?
   a. light energy
   b. potential energy
   c. kinetic energy
   d. free energy
   e. dark energy

Answer: d

Calls for factual knowledge.

3. One of the major differences separating prokaryotes and eukaryotes is the presence in eukaryotes of membrane-bound structures known as
   a. organelles.
   b. tissues.
   c. cytotoxins.
   d. nucleoids.
   e. systems.

Answer: a

A correct response requires conceptual understanding of these two cell types.

4. Which of the following structures prevents urine from leaking out of your bladder?

a. glycolipids
b. tight junctions
c. plasmodesmata
d. gap junctions
e. desmosomes

Answer: b

Factual knowledge required.

5. Which of the following structures could be described as "large, round, and membrane-bound"?

a. ribosomes
b. Golgi apparatus
c. chromosomes
d. endoplasmic reticulum
e. nucleus

Answer: e

Requires factual knowledge of the structure of the cell.

6. Which of the following structures could play an important role in the phenomenon known as "drug tolerance" in liver cells?

a. ribosomes
b. Golgi apparatus
c. chromosomes
d. endoplasmic reticulum
e. nucleus

Answer: d

Knowledge of cells applied to a real-world situation.

7. organic compounds + oxygen ⟶ carbon dioxide + water + energy

In the equation above, what is a common term used to describe "organic compounds"?

a. hydrocarbons

b. food

c. isomers

d. functional groups

e. dinucleotides

Answer: b

This calls for a *conceptual* understanding of the function of food and how organisms use it to create energy.

8. Which of the following plants have naked (non-fruit-covered) seeds?

a. algae

b. mosses

c. ferns

d. gymnosperms

e. angiosperms

Answer: d

9. Which of the following make up the culinary delicacy known as "truffles"?

a. epiphytes

b. rhizobia

c. mycorrhizae

d. gymnosperms

e. "O" fries (large, with cheese)

Answer: c

Relates to real-world experience. You need to know what truffles are in order to classify them.

10. How many types of bacteria are identified as being involved in nitrogen fixation?

a. 15

b. 12

c. 9

d. 6

e. 3

Answer: e

11. Imagine that you have two spider plants: "Jay" planted in a stressful, nutrient-poor soil, and "Dave" in nutrient-rich well-watered loam. Which one is likely to reproduce by asexual cloning?

a. Jay
b. Dave
c. neither

Answer: b

Requires application of a fundamental principle to a real-world situation.

12. Gametogenesis (from sporocyte to egg or sperm) in plants requires the action of which of the following?

a. auxin transport
b. bulk flow
c. mitosis and meiosis
d. abscission
e. transpiration

Answer: c

A conceptual question that requires a solid understanding of gametogenesis.

13. A plant that has flowers with pure red petals is probably

a. not producing nectar
b. not native to Europe
c. not pollinated by birds
d. not an angiosperm
e. extinct

Answer: b

Requires application of facts learned in class to a real-world case.

14. Another name for the growth of vines around tree trunks is

a. phototropism
b. gravitropism
c. thigmotropism

Answer: c

Factual knowledge called for.

15. All of the following are systems that plants and animals have in common EXCEPT

a. central nervous system
b. circulatory (or transport) system
c. reproductive system
d. endocrine system
e. All of the above are common to plant and animals.

Answer: a

The student should understand the similarities and differences between plants and animals. This is a fundamental conceptual question.

16. Sperm cannot swim through the material forming a barrier across the cervical opening. What hormone creates holes that allow the sperm to enter the cervix?

a. testosterone
b. FSH
c. estrogen
d. progesterone
e. none of the above

Answer: c

17. How many eggs and sperm are required to create identical twins in humans?

a. 2 eggs; 2 sperm
b. 1 egg; 2 sperm
c. 2 eggs; 1 sperm
d. 1 egg; 1 sperm
e. That information has not been discovered yet.

Answer: d

18. A normally pigmented couple has several children. Both the husband and the wife come from families in which one of the parents is albino (a Mendelian recessive trait). What fraction of the offspring of this couple is expected to have normal pigmentation?

a. 1/4
b. 2/4
c. 3/4

d. 2/3
e. 1/8

Answer: c

Aa × Aa may yield AA, 2Aa, and aa; therefore, 3/4 of the offspring should have normal pigmentation.

19. Sheep can be either black or white in color. These colors are determined by a single allele. A white ram (male) and white ewe (female) produce a black lamb. Which color is dominant?
a. black
b. white
c. the alleles are codominant
d. the alleles are epistatic
e. cannot determine from the information given

Answer: b

This question and the next two require factual knowledge of the principles of genetics and the ability to apply them conceptually.

20. A woman with normal color vision whose father is color-blind has four sons. We know that her husband is normal. What is the expected ratio of normal vision to color blindness among these sons?
a. 1:1
b. 1:2
c. 3:4
d. 2:1
e. cannot determine from the information provided

Answer: a

21. A family has four children, one with blood type A, one with blood type AB, and one with blood type O. What are the genotypes of the parents?
a. BB, Bi
b. AA, ii
c. Ai, Bi
d. AB, ii
e. All of the above are possible.

Answer: c

22. In tomatoes, round fruit (O) is dominant to elongated (o), and smooth fruit skin (P) is dominant to fuzzy skin (p). A homozygous round, fuzzy plant is crossed to a homozygous elongated smooth-skin plant, and their offspring (F_1) are then test-crossed. The results among the progeny from the testcross were:

420 round, fuzzy:57 round, smooth:63 elongated, fuzzy:460 elongated, smooth.

Interpret these results.
a. Round and smooth are epistatic.
b. Round and smooth are codominant.
c. Round and smooth are linked.
d. Round and smooth are multiallelic.
e. Round and smooth are redundant.

Answer: c

First determine the expected ratios from the testcross if unlinked, then compare this result to what was observed. Determine which answer fits best.

23. The probability of tossing three coins simultaneously and obtaining three heads is:
a. 1/2
b. 1/4
c. 1/8
d. 1/16

Answer: c

A coin toss is an example of 50/50 probability; therefore multiply 1/2 three times: $1/2 \times 1/2 \times 1/2$. The product is 1/8.

24. All of the following are major differences between oogenesis and spermatogenesis EXCEPT
a. oogenesis includes the formation of polar bodies.
b. spermatogenesis is continuous throughout the male's reproductive years.
c. oogenesis has long "resting" periods, in contrast to spermatogenesis.
d. spermatogenesis produces gametic cells, while oogenesis produces somatic cells.
e. all of the above

Answer: d

Students should have the factual knowledge to contrast male and female reproduction processes.

25. The trait for "hairy ears" is a Y-linked recessive. Therefore, we would predict that

a. it could skip generations.
b. sons would always get it from their mothers.
c. daughters would inherit it in a 1:1 ratio.
d. it would be observed as a Barr body in somatic cells.
e. none of the above

Answer: e

Application of conceptual understanding required.

26. Which of the following agents could be responsible for mad cow disease AND the sequel to Jurassic Park?

a. retroviruses
b. DNA viruses
c. rabies virus
d. virions
e. prions

Answer: e

Calls for application of knowledge—knowing that prions are the causative agents of mad cow diseases—to real-world examples.

27. Cheetahs are endangered in the wild. Which of the following indicates that cloning may be their best hope of survival?

a. No female cheetahs now exist in the wild.
b. Research suggests that cheetahs may have a genetic heterozygosity of 0.07%.
c. Cheetah evolution has resulted in the gradual loss of reproductive structures.
d. Sheep and cheetahs surprisingly share much of the same genome.

Answer: b

Real-world understanding is required here. Given such knowledge, most of the possible answers are easily eliminated.

28. Fossils are found in ______ rocks.

a. sedentary
b. sequential
c. supplementary

d. sedimentary

Answer: d

29. Phytoestrogens

a. were evolved by plants to control the fertility of animal predators.
b. were evolved by plants as a means of plant-to-plant communication.
c. were evolved by plants as part of an internal control system.
d. are the cause of the delicious taste of "O" fries.
e. All of the above are true.

Answer: c

Conceptual understanding is required, as is knowledge of real-world applications.

30. Who are the major predators of plants?

a. bacteria
b. fungi
c. birds
d. sporophytes
e. insects

Answer: e

Broad conceptual knowledge required.

31. Allelopathy is

a. the production of secondary compounds that inhibit the growth of other plants.
b. seed dispersal by ants.
c. replacement of carbon dioxide by oxygen during the Calvin cycle.
d. adding new DNA to a plant species.
e. the study of plant diseases.

Answer: a

A pitfall of guessing is suggested by answer e. The "-pathy" suffix frequently refers to disease conditions.

32. What fetal structures are responsible for obtaining nutrients and oxygen from the mother?

a. the yolk sac
b. the fetal aorta
c. the allantois
d. the chorionic villi
e. none of the above

Answer: d

33. Which two bases are complementary to the nitrogenous base adenine?

a. uracil and cytosine
b. guanine and thymine
c. thymine and uracil
d. ribose and clearasil
e. first base and home plate

Answer: c

Note the throwaway responses. They make guesswork more worthwhile.

34. Where does the process of transcription take place in the cell?

a. on the smooth endoplasmic reticulum
b. on the ribosomes
c. within the Golgi apparatus
d. within the nucleus
e. "on top of Old Smokey . . ."

Answer: d

Requires a conceptual understanding of cells.

35. Which of the following may be required for viral gene expression but NOT bacterial gene expression?

a. totipotential cells
b. operons
c. prions
d. the lytic cycle
e. design classes at Wrangler Co.

Answer: d

Calls for a conceptual understanding of genetics in viruses and bacteria. Most students should be able to derive the answer from eliminating the incorrect alternatives.

FINAL EXAM

Instead of saying "Study hard!" I would suggest "Study often!" I caution students that the synthesis stage of comprehension cannot be acquired during an all-nighter right before the exam. Reviewing and refining lecture notes should be done weekly, if not daily. I find that the good students seem to do this without prompting, while the truly poor students never get the point.

All questions in these exams are equally weighted; however, I hope that well-prepared students will answer the factual questions quickly and, therefore, have more time and energy to devote to the conceptual or problem-solving questions.

1. Extremophiles (organisms capable of thriving in extreme environments) are characteristic of which of the following groups?

a. archaebacteria
b. eubacteria
c. chordata
d. arthropods
e. bryophytes

Answer: a

Requires conceptual understanding of archaebacteria as anaerobes.

2. All of the following are used to classify the major divisions of plants EXCEPT

a. vascular system
b. seeds
c. needles
d. flowers
e. all of the above

Answer: c

Conceptual understanding of plants and of systematics is useful here.

3. You are standing in a SAVANNAH. By traveling in what direction could you reasonably expect to find TAIGA?

a. south
b. north
c. east
d. west

Answer: b

Another question calling for conceptual understanding, this time concerning biomes.

4. Pollution of the Earth has reached disastrous levels. As World Premier you must choose a group to travel on the spaceship *Noah* to colonize other planets and preserve the species. Which human subpopulation would you choose?

a. one with more K-selected adaptations
b. one with more r-selected adaptations
c. one with more Q-selected adaptations
d. To heck with them, I'm on the first ship out of here!

Answer: b

Calls for the ability to apply factual knowledge to a real-world situation.

5. Why are big, fierce animals rare?

a. limited number of potential mates
b. limited ability for finding "O" fries
c. limited availability of water
d. limited availability of sunlight from previous tropic latitude
e. limited availability of energy from previous trophic level

Answer: e

A correct answer requires a sound conceptual understanding of the food web.

6. Why are dogs "stupid"?

a. lack of imprinting
b. lack of habituation
c. lack of proximate causes of behavior
d. lack of insight
e. What can you say? They are just not cats . . .

Answer: d

Requires understanding of the concept of insight as the opposite of "stupidity." Make an effort to interpret the language of the question. Note that *insight* is more nearly the opposite of *stupid* than any of the other choices (imprinting and habituation) are. This is a good example of finding the answer *in* the question.

7. In the analogy of the immune system used in class, which of the following is the best description of the role played by the skin and mucous membranes?

a. the maiden in the tower
b. the stables
c. the throne room
d. the moat and walls
e. the knights and sentries

Answer: d

This conceptual question requires an ability to visualize analogy and to recall material covered in class.

8. Which of the following substances is produced by pathogens and causes an immune response?

a. inactivated vaccines
b. antigens
c. antibodies
d. eosinophiles
e. interferons

Answer: b

You need to know what each of these substances is.

9. What term would Darwin use to describe the following? "The whole purpose of such *mano a mano* combat is to secure breeding rights to females."

a. kin selection
b. biogeography
c. female choice
d. inclusive fitness
e. male-male competition

Answer: e

Conceptual understanding required. In multiple-choice questions, very often the obvious alternative is indeed the correct one. Nor do you have to imagine yourself as Darwin to make the right answer.

10. The major outcome of Mendel's work was to

a. explain the gaps in the fossil record.
b. demonstrate a possible origin of life on earth.
c. uncover the particulate mode of inheritance.
d. prove the endosymbiont theory.
e. unite "O" fries with cheese.

Answer: c

Requires conceptual understanding of the significance of Gregor Mendel's pioneering work in genetics.

11. Meiosis allows species to increase the genetic variation within the gene pool by

a. making exact copies of each parent cell.
b. using a diploid cell to make two haploid cells.
c. causing internal fertilization.
d. generating new cells to repair injury.
e. preventing cancer.

Answer: b

This combines factual knowledge with the application of the concept of genetic variation.

12. Which of the following would be the major difference between the cell cycles of a human nerve cell and an *E. coli* (bacterial) cell?

a. All cells have identical cycles.
b. The neuron does not have a cell cycle.
c. The *E. coli* cell remains in the S phase for extended periods.
d. *E. coli* divide only by meiosis.
e. The neuron remains in the G phase.

Answer: e

A successful answer requires understanding key cellular concepts.

13. Darwin's work confirmed

a. evolution through acquired characteristics.
b. that the cell is the basic unit of life.
c. the molecular basis of inheritance.
d. the unity and diversity of life.

e. the endosymbiont theory.

Answer: d

The question tests the student's grasp of the significance of Darwin's work. It calls for conceptual understanding.

14. Human beings, as a species, are capable of eating plant and animal foods and are

a. herbivores.

b. autotrophs.

c. carnivores.

d. omnivores.

e. none of the above

Answer: d

It is important to know the definitions of key terms.

15. ANIMALS with more than one cell are called

a. protozoans

b. metazoans

c. multitropic

d. heterotrophs

e. autotrophs

Answer: b

See comment for 14.

16. Energy is neither ______ nor ______.

a. transformed, transduced

b. processed, perturbed

c. created, destroyed

d. evolved, hydrolyzed

e. created, transduced

Answer: c

Requires knowledge of the concept of conservation of energy.

17. The situation where two or more species have similar structures used for the same

purposes even though the species are not closely related is known as

a. convergent evolution.
b. divergent evolution.
c. retrospective evolution.

Answer: a

Tests understanding of basic evolutionary concepts.

18. A process in which humans select which plants and animals shall live and die is called

a. natural selection.
b. organic selection.
c. artificial selection.
d. artificial insemination.

Answer: c

A basic understanding of natural selection will eliminate option *a*, and none of the others make sense except *c*. Requires understanding the concept of natural selection.

19. Lamarckian theory suggests the inheritance of ______ characteristics, while Mendelian theory proposes the inheritance of ______ characteristics.

a. fit, reproductive
b. diversifying, stabilizing
c. acquired, intrinsic
d. convergent, divergent

Answer: c

Requires factual and conceptual knowledge of Lamarckian theory versus Darwinian theory.

20. Creationism is not a science because

a. its hypotheses are not testable.
b. it does not require the use of beakers and test tubes.
c. the secret societies of scientists refuse to accept it.
d. All of the above are true.

Answer: a

Although the answer is readily derived by process of elimination, the question does call for understanding a concept basic to the scientific method: a hypothesis must be falsifiable—testable.

21. Which of the following is accepted by scientists to be a core concept of biology?

a. spontaneous generation
b. creationism
c. Third Law of Thermodynamics
d. evolution
e. ecology

Answer: d

Requires understanding of what a "core concept" is.

22. Why are the bottoms of begonia leaves red?

a. to enhance photosynthesis
b. as an antipredator device
c. to attract pollinators
d. as an allelopathic device
e. because it looks pretty with all that green

Answer: a

Calls for a combination of factual knowledge and a conceptual understanding of the photosynthetic process.

23. Which of the following is a valid observation about the coevolution of angiosperms and their pollinators?

a. Solitary bee species pollinate different species of plants.
b. Moth-pollinated flowers bloom only in the morning.
c. Most bird pollinators are native to Europe.
d. The effective range of wind pollination is 100 km.
e. Insects are sensitive to color and therefore only pollinate red flowers.

Answer: a

Requires conceptual understanding.

24. Which of the following would be considered an aspect of the dual meaning of Darwinism?

a. Darwinism introduced independent assortment and segregation.
b. The taxonomic work of Linnaeus was proven false by Darwinism.
c. Evolution explains the unity and diversity of life.

d. Darwinism explains both Lamarckian evolution and the modern synthesis.
e. Natural selection affects only fossil (nonliving) species.

Answer: c

The question tests conceptual understanding of the broader significance of Darwinian theory.

25. What is the purpose of systematics?

a. All of those below are correct.
b. to reveal natural law
c. to perform experiments in sequence so as to quantify results
d. to connect biogeography and continental drift
e. to relate biological diversity and phylogeny

Answer: e

A successful response requires understanding the concept of systematics. Most students will have little trouble eliminating the incorrect responses.

26. What term describes the pressures that lead to evolution?

a. tectonic pressure
b. independent assortment
c. scala naturae
d. Polyploidy
e. natural selection

Answer: e

Another question testing conceptual understanding.

27. Compare and contrast divergent and convergent evolution:
Which of the following would result in a group of species that are physically similar but NOT genetically related?

a. divergent evolution
b. convergent evolution
c. both would have that result
d. neither would have that result
e. one of the above, but I don't know which

Answer: b

Requires understanding the concepts of convergent and divergent evolution.

CHAPTER 24

SAN JOSE STATE UNIVERSITY

BIOLOGY 20: ECOLOGICAL BIOLOGY

Michael Kutilek, Assistant Professor

THREE EXAMS ARE GIVEN IN BIOLOGY 20, INCLUDING TWO MIDTERMS AND A FINAL. All are essay-type exams, and together they account for 75 percent of the course grade. A "midterm" and final are presented here.

The main goal of the course is to introduce nonscience majors to general principles of ecological and evolutionary biology, and to build awareness about environmental issues, especially the conservation of biodiversity. I want students to build their critical-thinking, problem-solving, and communication skills. I want them to understand the workings of the scientific method, its power and its limitations. I also want them to have a broad knowledge of ecology and evolution, so that they are able to evaluate environmental/biodiversity problems as scientific issues and as pressing societal problems.

My students write essay exams in which they are required to use critical-thinking, problem-solving, and communication skills. Rote memorization is of limited value because they are asked to integrate what they have learned into larger issues of how modern human beings interact (or should interact) with nature. In some cases they are asked to apply what they have learned to solve problems. The students need to have a reasonably deep understanding to function at this level.

Students should attend every class, be on time, stay alert, take complete notes, participate fully in the discussions, ask questions as they come up, get extra help by taking advantage of office hours, form small study groups, and ask each other penetrating questions. Memorization should be minimal and a minor concern; however, students should study as they go and not let themselves get behind in the course.

MIDTERM EXAM (EXAM I)

There are two comprehensive questions on this exam. Before you begin to write, spend time thinking about these questions and developing detailed outlines. Be logical, concise, and complete in your response. You will be graded not only on what you say, but also on how well you say it; therefore, use proper grammar, sentence structure, etc., and avoid redundancies.

Question 1

The Wampus is a large grazing antelope that lives on the grasslands of the planet Macroevol. The fossil record of this animal shows that over the past 20,000 years, the average member of this species has increased in size significantly; shown increased development of a single, pointed horn in the middle of its forehead; and an eyelike structure in the rear of its head that is able to detect motion. The Wampus is preyed upon exclusively by the Woofapoof, a predator whose fossil record shows that it too has increased in size and has taken to hunting in groups. Woofapoofs most often attack a Wampus either by charging it head on or by sneaking up on it from behind. We know that the forces of natural selection operate the same way on Macroevol as they do on Earth. What specifically are the *major forces of natural selection* on the Wampus and on the Woofapoof? Using terms like *survival, reproduction, species, genotype*, and *phenotype*, develop an explanation of how these changes have occurred over time that is consistent with the *process* of natural selection.

> **You should not begin writing immediately, but should spend time understanding the question and framing a thoughtful response by developing an outline. Arrange your outline so that it makes good logical sense. Edit your response for grammar and flow so that the final product is factually correct, logically plausible, and error-free. The final product should be university-level work in style and content.**

Answer: The major force of natural selection on the Wampus is predation by the Woofapoof, while the major force of natural selection on the Woofapoof is the set of evasive/protective activities by the Wampus. This is a clear example of coevolution, where the activities of each species have acted as a strong selective force on the other.

In any large natural population, there are many varied phenotypes. In the case of the Wampus, some were larger in size, some smaller; some had a bony bump on their head (precursor to a horn), some did not; some had sensory tissue on the back of the skull (precursor to an eyelike sensor), some were without. The characteristics displayed by phenotypes have an underlying genetic basis, which can be passed from parents to offspring, i.e., each phenotype has an underlying genotype. Selection operates on phenotypes. In the case of the Wampus, the phenotypes that had the best survival and reproduction were those that were better able to detect and fight off predators. Therefore, early Wampus phenotypes that showed a tendency toward being larger in size, having the beginnings of a horn, and having an eyelike detector in the back of the head, became more prominent with each succeeding generation.

There were also many varied Woofapoof phenotypes, but Woofapoofs that had the greatest survival and reproduction were those that could best overcome the defenses of the Wampus. Woofapoofs that were larger in size and that took to hunting in groups could more readily overpower the prey and became prominent as their genes were passed along. Coevolution is ongoing and operates through many small incremental changes over a long period of time.

The answer is effective because it addresses each of the issues asked for in the question. The student has been careful to use in his or her response all of the key terms called for in the question. Most of all, this response demonstrates an understanding of principles that is sufficiently thorough to allow the student to apply them to a specific set of circumstances.

Question 2

Species diversity is a major area of study in ecology. What is species diversity? Is there a possible relationship between species diversity and the stability of communities? Describe a prominent global pattern of species diversity. Discuss the factors that are likely to have caused differences in species diversity on a local and global scale. Briefly describe one natural community that has low species diversity and another with high species diversity, using examples we have seen on one of our field trips.

Answer: Species diversity is a subcategory of biodiversity that refers to the number of species that make up a community. Species diversity may aid in stability, especially when a community is attacked by a force such as disease; the mix of species in the community may prevent the disease from wiping out all members of one or more species. A prominent global pattern of species diversity is the general decline in the number of different species as one moves from the equator to the poles. The factors that are likely to have caused differences in species diversity on a global and local scale include

(1) Time and Environmental Stability: For example, that the tropics have had more time free of massive disturbance (like glaciation) and therefore more time for species to evolve.

(2) Spatial Diversity: Habitats that have more diversity in both horizontal and vertical dimensions (forests, ecotones) have more niches and, therefore, more species to occupy them.

(3) Interplay Between Competition and Predation: The activities of predators reduce competition among prey species and, therefore, allow more prey species to coexist.

(4) Stability of Productivity: Areas like rain forests that have year-round, reliable food resources will have more resident species that live there and are specialized to use those resources.

(5) Intermittent, Intermediate Levels of Disturbance: Intermediate disturbance (fires, floods, windstorms) set back succession and create a mix of many successional stages over the landscape, which in turn give rise to maximum species diversity.

Examples of communities we have visited with high species diversity include mixed woodland, chaparral, riparian, and aquatic habitats. Those with low species diversity include salt marshes and salt ponds.

This question permits a good deal of leeway, but that does not mean that a poorly presented or poorly reasoned answer is adequate. As with the first question, this response is successful in large part because the student has taken care to address each issue mentioned in the question. In contrast to the first question, however, more narrowly factual information is called for. Appropriately, therefore, the student uses a list format for part of his or her response.

FINAL EXAM

I would stress that the students need to work toward a deeper level of integration and understanding. Mere term recognition or memorization of definitions is not enough. They need to get to the point where they can examine a problem they have not directly encountered before, analyze it in light of what they know, and come to a reasonable conclusion.

There are two comprehensive questions on this exam. Before you begin to write, spend time thinking about these questions and developing detailed outlines. Be logical, concise, and complete in your response. You will be graded not only on what you say, but also on how well you say it; therefore, use proper grammar, sentence structure, etc., and avoid redundancies.

Question 1

You are acting as a consultant to the Ministry of Parks in Brazil as they consider their options for a 1000-square-mile tract of rain forest in the Amazon. The region is rich in biodiversity. Along with large areas of forest, there are also rivers, lakes, and patches of grasslands. The area is subject to windstorms that sometimes uproot trees and to periodic fires caused by lightning.

The Brazilian government is going to develop part of this area by harvesting some of the timber and by making permanent human settlements. They want to do this in a way that will give the settlers a good chance for long-term economic gain. They also want to practice conservation, establish national parks, and encourage tourism.

What advice do you give them? What overall philosophy do you try to impart? What specific short-term and long-term strategies do you suggest to them to maximize the economic return while at the same time conserving and managing biodiversity?

Answer: The overall philosophy should be to strive to maintain the integrity of the processes that gave rise to the forest. Recognize that some changes in species composition and density over time are natural. Natural catastrophes, succession, and natural selection are ongoing processes that give rise to change. Rather than trying to maintain a constant environment, one must allow these natural processes to continue. Some human changes are natural and appropriate as long as they fit within the functional, historical, and evolutionary limits of the system.

Take special notice of the opening sentence of this first paragraph. It establishes the main theme of the response as well as the guiding principle motivating the response. Such opening sentences are especially important in responding to essay questions because they guide the examiner's perception, interpretation, and evaluation of the rest of the response.

Functional limits are exceeded when anthropogenic changes go beyond the physiological tolerances of organisms in the system, historical limits are exceeded when exotics are introduced that interfere with natives, and evolutionary limits are exceeded when landscapes are altered at such a rapid rate that anthropogenic selection becomes dominant. Settlers should be trained to find, use, and market economic resources without destroying the forest. Such resources include foods, fiber, oils, medicines, and natural herbicides and pesticides. Clear-cutting must be avoided at all costs since destruction is irreversible, but some selective cutting is appropriate.

A thoughtful, concise paragraph. The student defines how natural limits may be violated, then goes on to make recommendations designed to avoid violating those limits. The paragraph ends by demonstrating recognition that certain extreme practices—clear-cutting—should be avoided (and explains why they should be avoided), but concedes that less extreme forms of the practice ("selective cutting") are appropriate. In short, the response is reasoned rather than extreme or dogmatic.

Developing ecotourism as a means of employment and a source of revenue is a definite possibility.

Note that the question mentioned only "tourism," but that the student uses the word "ecotourism." Always take advantage of opportunities to use precisely the correct term. Demonstrating a command of appropriate terminology is key to performing well on essay exams, especially in subjects that draw heavily on specialized vocabularies.

Question 2

Write a clear, complete, and coherent essay on human nutrition. What are all the components of a good diet ("diet" defined in broad terms to mean any substances we take into our bodies)? How can an individual develop the best diet for himself or herself? How can one have the best probability of ensuring a long, healthy, and disease-free life through good nutrition? What are the disease risks of a poor diet? How does one's choice of diet affect conservation of the earth's natural resources?

It would be very easy to become overwhelmed by this question. However, read the question carefully. It provides its own set of specifications for answering it. Confronted with a multipart question, it is usually most effective to use the parts as the structure of your outline. The instructor expects an orderly response to each of the questions asked. Deliver amply on this expectation.

Answer: Components of a good diet include carbohydrates, proteins, fats, vitamins, minerals, and fiber. One should take in 50-65% of calories as carbos, 10-15% as proteins, 15-30% as fats (unsaturated fats are best), and vitamins, minerals, and fiber in smaller amounts or as trace elements. The diet should include a rich mix of fresh fruits and vegetables to ensure appropriate nutrition without excessive caloric intake. One should not consistently take in more calories than can be metabolized per day or obesity will result.

The response shows a solid command of factual knowledge.

Poor diet is correlated with a higher risk of cardiovascular disease and cancer. To lower disease risk, reduce dietary fat below 30% of calories, reduce cholesterol (to help prevent heart disease), increase fiber intake to >20-30 g/day (to help move metabolic wastes through the colon), increase intake of beta-carotenes (in green and orange veggies), vitamin C (citrus), and cruciferous veggies (broccoli et al.); the latter three all have protective qualities (they are antioxidants).

The preceding paragraph is effective because it concisely defines the health risks of a poor diet and the health benefits of a good one. Note that the student is careful about overasserting cause and effect. He uses the phrase "is correlated with" rather than "causes." This reveals a sound understanding of how scientists interpret data and draw conclusions from it.

Diets rich in meats, especially red meat, may not only be unhealthy, but are environmentally unsound; that is, they use more resources lower in the food web to support herds of cattle. The meat industry also uses large amounts of water, energy, and other resources.

The concluding paragraph effectively answers the last part of the question, broadening the response from the individual to the ecosystem. Note that the response is not expressed in emotional or "political" terms, but in emotionally neutral terms of impact on the ecosystem.

CHAPTER 25

SAN JOSE STATE UNIVERSITY

BIOLOGY 21: HUMAN BIOLOGY

Elizabeth McGee, Assistant Professor

NO MIDTERM OR FINAL IS GIVEN IN THIS COURSE. INSTEAD, STUDENTS COMPLETE fourteen weekly exams, eight of which are presented here.

Primary course objectives are to provide students with a general overview of key issues in human biology, to increase science literacy, and to encourage lifelong learning of science. I want my students to internalize information acquired in class in order to make informed decisions about personal health issues and environmental issues, and to be informed voting citizens.

How do the exams relate to the course? Several questions are based on essays that appear in the course textbook (Sylvia Mader's *Human Biology*). I do not test trivial detail. This is important, because it makes the course appear less intimidating and more inviting.

Success in this course requires that students:

- show up
- read
- learn to "concept map" the chapters to pick out the relevant information

Concept mapping requires summarizing the essays in the textbook. This will help prepare for short-answer questions on the exams. Students should also concentrate on vocabulary at the end of each textbook chapter: know what the terms mean and how they relate to the topic of the week. Realize that everything has a purpose: know what it is. Students don't *naturally* want to know how biological systems function, and they often struggle as a result.

We give students a "need-to-know" list in lecture every week and tell them to read the chapter for the items on this list. We also tell them what we *don't* think is relevant or important for a course in general human biology.

WEEK 1 EXAM: HUMAN EVOLUTION

Mutiple Choice *(Select the one best answer for the following.) (1 point)*

1. The structural similarities between the wings of a bat and forelimbs of a frog are an example of

a. analogies.
b. embryological remnants.
c. homologous structures.
d. vestigial structures.

Answer: c

2. The name of the species to which we belong is

a. *Homo neanderthalensis.*
b. *Homo erectus.*
c. *Homo humanensis.*
d. *Homo sapiens.*

Answer: d

3. Which of the following areas of study provides evidence to support evolution?

a. comparative biochemistry
b. comparative anatomy
c. fossils
d. all of the above

Answer: d

4. Which of the following species is considered the most ancient?

a. *Australopithecus afarensis*
b. *Homo habilis*
c. *Homo erectus*
d. *Australopithecus boisei*

Answer: a

5. Which of the following hominids made tools?

a. *Homo habilis*

b. *Homo sapiens*
c. *Homo erectus*
d. all of the above

Answer: d

6. Which of the following is not an anthropoid?

a. lemur
b. baboon
c. gorilla
d. golden-lion tamarin

Answer: a

7. The earliest primates evolved in what type of habitat?

a. savanna grasslands
b. tropical forests
c. arctic tundra
d. deltaic swamps

Answer: b

8. Because humans have a backbone, they are members of what phylum?

a. Arthropoda
b. Chordata
c. Hominidae
d. Mammalia

Answer: b

9. As *Homo* evolved, certain traits and physical characteristics became associated with more advanced species. Which is NOT considered an advancement?

a. bipedal locomotion
b. flattened face
c. larger brain
d. larger brow ridge

Answer: d

10. Which of the following is NOT part of Darwin's theory of natural selection?

a. Natural selection results in the adaptation of populations to their specific environments.
b. Acquired characteristics of parents will be inherited by their offspring.

c. There are inheritable variations among members of a population.
d. Many more individuals are produced each generation than can survive and reproduce.
e. Individuals with inheritable adaptive characteristics are more likely to survive and reproduce.

Answer: b

11. Which of the following is not a Kingdom?
a. Protista
b. Animalia
c. Plantae
d. Mammalia
e. Fungi

Answer: d

12. The earliest hominids are approximately how old?
a. 2 million years
b. 3.5 billion years
c. 4 million years
d. 500,000 years

Answer: c

True/False *(1 point)*

13. Chimps are ancestral to humans.
Answer: F

14. Cro-Magnon man belongs to the species *Homo sapiens.*
Answer: T

15. "Races" are defined as populations within a species that cannot interbreed.
Answer: F

16. *Homo habilis* used fire to make tools from iron.
Answer: F

17. A fossil is any remains of an organism that have been preserved in the earth's crust.
Answer: T

18. Sense of smell is the most important sense in primates.
Answer: F

19. The term "evolution" means change over time.
Answer: T

20. Plants were probably the first living organisms on land.
Answer: T

Short Answer

21. Name three characteristics of primates that make them especially well-adapted for living in trees. Do not just list characters; describe what they are and why they are important to tree-dwellers. (6 points)
Answer:
1. grasping hands and feet (easier to grip branches)
2. nails associated with tactile pads and dermatoglyphs (since primates don't have claws, the pads and fingerprints make it easier to grip without slipping)
3. single births (easier to carry one offspring than many)
4. visual "suite": sharper vision, color vision, and depth perception (you need excellent vision when living in an environment where you have to gauge distance between branches, etc.)

22. Discuss three lines of evidence for bipedality in the early australopithecines. Do not just list characters; describe what they are. (6 points)
Answer:
1. skull: position of the skull on the vertebral column (vertebral column is directly underneath skull)
2. hips: iliac blades of pelvis are short and broad
3. knees: the articulating surface between the femur and tibia is relatively flat; allows us to lock our knees and conserve energy; also thigh bones angle in (valgus knee)
4. feet: big heel bone; big toe is in line with rest of foot; this stabilizes the foot and allows efficient push off and heel strike of the foot

Bonus: An interesting article appeared in the *San Francisco Chronicle* this week about a project at the San Francisco Zoo to reintroduce a group of primates back into their native habitat. What were these animals? (1 point)
Answer: ruffed lemurs

Bonus: What is the term for "living in trees"? (1 point)
Answer: arboreal

WEEK 2 EXAM: ORGANIZATION OF LIFE

Multiple Choice *(Select the one best answer for the following.) (1 point)*

1. Which type of animal tissue covers body surfaces and lines body cavities?
a. nervous
b. connective
c. epithelial
d. muscular
Answer: c

2. Amino acids are the building blocks of
a. proteins.
b. carbohydrates.
c. lipids.
d. DNA.
Answer: a

3. Regarding the article on dinosaurs that appeared in the January 27 issue of the *San Jose Mercury*, what country or region were the dinosaurs mentioned in this article found in?
a. Portugal
b. Tasmania
c. Pennsylvania
d. Romania
Answer: d

4. Also regarding the article on dinosaurs that appeared in the January 27 issue of the *San Jose Mercury*, what was unusual about these dinosaurs?
a. they were the oldest dinosaurs discovered to date
b. they evolved on an island
c. they were exquisitely preserved
d. all of the above
Answer: b

5. The function of the endoplasmic reticulum is to

a. control the cell.
b. digest macromolecules.
c. act as a transport system.
d. move the cell.

Answer: c

6. Which of the following is a polysaccharide?

a. sucrose
b. lactose
c. glucose
d. cellulose

Answer: d

7. The Golgi body is involved with

a. packaging and secretion.
b. cell reproduction.
c. making proteins.
d. all of the above

Answer: a

8. Which of the following are characteristics of UNSATURATED fat?

a. found in foods derived from animals and are solid at room temperature
b. found in foods derived from animals and are liquid at room temperature
c. found mainly in plants and are solid at room temperature
d. found mainly in plants and are liquid at room temperature

Answer: d

9. Humans temporarily store excess carbohydrates as

a. glucose.
b. starch.
c. glycerol.
d. glycogen.

Answer: d

10. The three types of muscle tissue are

a. cardiac, smooth, cartilage.
b. cardiac, skeletal, cartilage.
c. nervous, skeletal, cardiac.

d. nervous, skeletal, smooth.
e. cardiac, skeletal, smooth.

Answer: e

11. Autotrophs

a. give off oxygen.
b. include plants.
c. carry on photosynthesis.
d. obtain their energy from the sun.
e. all of the above

Answer: e

12. All of the following are macromolecules except

a. proteins.
b. carbohydrates.
c. lipids.
d. b and c.
e. All of the above can be macromolecules.

Answer: e

13. Protein synthesis is carried out at which of the following organelles?

a. lysosomes
b. nucleus
c. ribosomes
d. nucleolus
e. mitochondria

Answer: c

True/False *(1 point)*

14. Inorganic molecules contain carbon, while organic molecules do not.
Answer: F

15. Anaerobic respiration uses oxygen.
Answer: F

16. Both DNA and RNA are nucleic acids.
Answer: T

17. Compound molecules are the basic units of organic and inorganic life.
Answer: F

18. Fat, cartilage, blood, nervous tissue, and bone are all kinds of epithelial tissue.
Answer: F

19. Mitochondria are the powerhouses of cells.
Answer: T

20. Lysosomes contain enzymes that digest macromolecules.
Answer: T

Short Answer: *Answer one of the following questions. If you do both, we will read only the first one!*

21. "Health Effects of Fiber" (10 points)

A. What is fiber? Give some examples of foods containing fiber. (2 points)
B. Give FOUR examples of how fiber promotes health. For each example, tell us what the benefit is, and specifically how it is achieved. (8 points)
C. BONUS: What is the difference between soluble and insoluble fiber?

Answers:
A. Fiber is a substance that cannot be broken down by human digestive enzymes; examples include wheat bran, wheat products, brown rice, cooked lentils, navy beans, green beans, kidney beans, oat bran, carrots, barley, and apples.

B.

Students should list four out of the eight actually listed below. One point is awarded for listing the item, and one point for describing a little bit about how it works.

1. weight control
2. constipation and diarrhea relief
3. hemorrhoid prevention
4. appendicitis prevention
5. diverticulosis prevention
6. blood lipid and cardiovascular disease control
7. blood glucose and insulin modulation
8. diabetes control

C. BONUS: Soluble and insoluble fiber function differently in the digestive tract. Soluble fiber, such as oat bran, soaks up bile acids and causes the liver to break down more cholesterol in order to make more bile acids. Soluble fiber, therefore, lowers blood cholesterol and prevents heart disease.

Insoluble fiber, such as wheat bran, provides bulk and speeds elimination of feces. Regular elimination reduces the time cancer-causing substances such as bile acids are in contact with the intestinal lining. Insoluble fiber, therefore, maintains regularity and prevents colon cancer. Insoluble fiber also reduces the absorption of dietary calcium, which may also be protective against colon cancer.

This is a good example of a tidy extended short answer. Note that the first sentence is a clear thesis sentence, which precisely identifies in what area soluble and insoluble fiber differ. The first paragraph then goes on to develop a concise explanation of the role of soluble fiber. Note the use of the word "therefore" in the last sentence of the first paragraph. This is a verbal signpost pointing out a conclusion from facts.

The second paragraph is devoted to insoluble fiber, and this is clearly announced in the first sentence of the paragraph. The main benefit of insoluble fiber is speeding the elimination of feces. This is explained thoroughly, with its benefit (prevention of colon cancer) clearly stated and explained.

22. "Start Saving Your Skin" (10 points)

A. What is the scientific term for skin cancer? Name the two general categories of skin cancer. (2 points)

B. Give FOUR examples of how you can prevent skin cancer. Be very specific! (8 points)

C. BONUS: What is the difference between UV-A rays and UV-B rays in the way they affect the skin? (2 points)

Answers:

A. Melanoma. Melanoma and Nonmelanoma.

B. Prevention

1. Use sunscreen; pick a broad-spectrum sunscreen (one that protects against UV-A and UV-B) with an SPF of at least 15.
2. Wear protective clothing; choose fabrics with tight weave and wear a brimmed hat.
3. Stay out of the sun; best hours are between 10 a.m. and 3 p.m.
4. Wear sunglasses treated to absorb UV-A and UV-B; pupils dilate more in the shade, thereby exposing eyes to more damage than usual.
5. Avoid tanning machines; although most use levels of UV-A, the deep layers of the skin become more vulnerable to UV-B radiation when you're later exposed to the sun.

You could also say "Move to Anchorage (or some other high-latitude area)," which is theoretically correct. Higher latitudes have lower UV indexes.

Take a look at the structure and content of each of these answers. The student avoids two- or three-word answers, stating instead the preventive step that may be taken, followed by a concise explanation of the benefits of the step. This is a good example of how to give short-answer responses impressive depth.

C. BONUS: UV-A rays penetrate the skin deeply, affect connective tissue, and cause the skin to sag and wrinkle. UV-B are cancer-causing rays.

Concise and to the point, this response would get full bonus credit. If, however, this were a *main* question, a sentence or two on the mechanism by UV rays act on skin would be appropriate.

WEEK 7 EXAM: GENETICS

To answer questions 1-4, use this key:

a. heterotype
b. genotype
c. heterozygous
d. phenotype
e. homozygous

1. The outward appearance of an organism

2. Having two different alleles for a given trait

3. Having identical alleles for a given trait

4. The genetic composition in an individual for a particular trait

Answers:

1. d
2. c
3. e
4. b

To answer questions 5-10, use this key (you may use terms in this key more than once):

a. sex-linked inheritance
b. sex-influenced inheritance

c. autosomal inheritance
d. chromosomal inheritance

5. color-blindness

6. hair and eye color

7. Huntington's disease

8. baldness

9. cystic fibrosis

10. hemophilia

Answers:
5. a
6. c
7. c
8. b
9. c
10. a

To answer questions 11-15, use this key (you may use terms in this key more than once):

a. mitosis
b. meiosis
c. both meiosis and mitosis

11. occurs in the reproductive organs

12. involves spindle fibers

13. increases genetic variation

14. occurs during growth and repair

15. chromosomes are cut from diploid to haploid

Answers
11. b
12. c
13. b
14. a
15. b

True/False *(1 point)*

16. In humans, the female determines the sex of the offspring.
Answer: F

17. The older a mother, the more likely she is to have a baby with Down syndrome.
Answer: T

18. It is impossible for a male to inherit an allele for an X-linked trait from his mother.
Answer: F

19. Alternate forms of a gene that influence the same characteristic and are found at the same location in homologous chromosomes are called alleles.
Answer: T

20. If a person's blood phenotype is A, then his blood genotype can be either AA or OO.
Answer: F

21. Humans have 48 chromosomes occurring in 24 pairs.
Answer: F

22. Skin color is a polygenic trait.
Answer: T

Punnett Problem *(8 points)*

A man is heterozygous for a widow's peak hairline; a woman is homozygous for a continuous hairline. Widow's peak is dominant (W) and continuous hairline is recessive (w).

What is the man's genotype for hairline?
Answer: Ww

What is the woman's genotype for hairline?
Answer: ww

Copy these answers to the following Punnett square (i.e., the lines below the father and the lines to the right of the mother). Then calculate the possible genotypes of their children.

	Father W	w
Mother w	Ww	ww
w	Ww	ww

Theoretically, if this couple has four children, how many will have a widow's peak?
Answer: 2

How many will have a continuous hairline?
Answer: 2

San Jose Mercury **Bonus Question** *(3 points)*

What is the name of the plant that produces cocoa beans? Where did this plant evolve? How could worldwide demand for chocolate help preserve rain forests?
Answer: The name of the plant is *Theobroma cacao*, and it evolved in South and Central America. Cacao plants are worth more alive than dead. This could halt the cutting down of rain forests.

WEEK 8 EXAM: MOLECULAR GENETICS AND CANCER

Multiple Choice *(Select the one best answer for the following.) (1 point)*

1. The structure in DNA that contains the source of coded information is the
a. sugar/phosphate backbone.
b. nitrogen bases and their sequences.
c. deoxyribose sequence.
d. ribose sequence.
e. all of the above
Answer: b

2. Mutations are the direct result of changes in
a. protein function.
b. phenotype.
c. amino acids.
d. nucleotide sequence.
Answer: d

3. In recombinant DNA research, the plasmids are taken from

a. plant cells.
b. animal cells.
c. bacterial cells.
d. any kind of cell.

Answer: c

4. Transcription occurs in the

a. cytoplasm.
b. ribosome.
c. nucleus.
d. mitochondrion.

Answer: c

5. The sequence of a DNA segment that was coded for a piece of mRNA that has the sequence UCACGU was

a. TTGGCA
b. ATCCGA
c. AGTGCA
d. AGUGCA

Answer: c

6. Translation takes place

a. in the nucleus.
b. on free ribosomes.
c. in the rough endoplasmic reticulum.
d. a and b are correct.
e. b and c are correct.

Answer: e

It is important to know the difference between *transcription* and *translation* in molecular genetics.

7. ______ cancer causes the most cancer-related deaths in humans.

a. colon
b. breast
c. lung
d. stomach

Answer: c

8. The two primary functions of DNA are _________ and _________.

a. transcription/translation
b. transcription/replication
c. replication/protein synthesis
d. replication/translation
e. transcription/protein synthesis

Answer: c

9. Cancer of the cervix is best diagnosed by

a. mammography.
b. X-rays.
c. ultrasound.
d. a Pap smear.

Answer: d

To answer questions 10-13, use this key:

a. protein
b. DNA
c. amino acids
d. tRNA
e. mRNA

10. codon

11. anticodon

12. sequence of bases in threes

13. transported to ribosomes by tRNAs

Answers

10. e
11. d
12. b
13. c

To answer questions 14-18, use this key:

a. carcinoma
b. lymphoma
c. adenocarcinoma
d. leukemia
e. sarcoma

14. Cancers of the blood

15. Cancers of the epithelial tissue

16. Tumors of lymphoid tissue

17. Cancers of the glandular epithelial cells

18. Cancers of the muscles and connective tissues

Answers
14. d
15. a
16. b
17. c
18. e

True/False *(1 point)*

19. DNA consists of ribose sugar, a phosphate, and four nitrogenous bases.
Answer: F

20. A fecal occult blood test is used to detect leukemia.
Answer: F

21. Human papillomavirus has been linked to cancer of the cervix.
Answer: T

22. mRNA is a double helix.
Answer: F

23. Cigarette smoke contains many chemical carcinogens and is associated with one-third of all cancers.
Answer: T

24. RNA bases include adenine, thymine, guanine, and cytosine.
Answer: F

25. Chemotherapy uses drugs to selectively kill off cancer cells.
Answer: T

26. Vitamins B and D are associated with the avoidance of cancer.
Answer: F

27. An oncogene is a gene that suppresses the growth of cancerous tumors.
Answer: F

28. Someone who both drinks and smokes is more likely to develop cancers of the mouth, larynx, bladder, kidney, and pancreas.
Answer: T

29. A bone marrow *autotransplant* involves harvesting bone marrow from a close relative.
Answer: F

30. DNA toenail painting can identify particular individuals in forensic and/or criminal cases.
Answer: F

WEEK 9 EXAM: CIRCULATORY SYSTEM

Multiple Choice *(Select the one best answer for the following.) (1 point each)*

1. In the systemic circuit, the heart's _________ half pumps ___________ blood to all body regions.
a. right; oxygen-poor
b. right; oxygen-rich
c. left; oxygen-poor
d. left; oxygen-rich
Answer: b

2. Blood pressure is highest in ________ and lowest in ________.
a. arteries; veins
b. arteries; ventricles
c. arteries; relaxed atria
d. arterioles; veins
Answer: a

3. Which of the following statements is correct?
a. A person with blood type B has the B antigens and anti-A antibodies, thus the person cannot receive type A blood.
b. Type O blood is considered a universal donor.
c. A person with blood type AB can only receive blood type AB.

d. all of the above
e. a and b only

Answer: e

4. Blood flow velocity is slowest in the
a. arteries.
b. arterioles.
c. capillaries.
d. venules.
e. veins.

Answer: c

5. Gas exchange takes place in the
a. arteries.
b. arterioles.
c. capillaries.
d. venules.
e. veins.

Answer: c

6. Arteries have ____________ and _________ walls when compared to veins.
a. thicker; stronger
b. thicker; weaker
c. thinner; stronger
d. thinner; weaker

Answer: a

7. How do nutrients and gases (i.e., oxygen) get to the heart muscle cells?
a. by diffusion from the blood in the atria and ventricles
b. by the pulmonary circuit
c. by the systemic circuit
d. by the coronary vessels
e. Heart muscle cells do not need nutrients or gases.

Answer: d

8. Movement of lymph within the lymphatic vessels is a result of
a. high blood pressure.
b. contraction of skeletal muscles.
c. contraction of smooth muscles found around the lymphatic vessels.

d. a and c only

Answer: b

9. Lymph (tissue fluid) is composed of

a. plasma.

b. large proteins.

c. formed elements.

d. all of them

e. a and c only

Answer: a

10. Which blood vessels have walls only one cell thick?

a. capillaries

b. venules

c. veins

d. arterioles

e. arteries

Answer: a

11. Which phase of the ECG indicates ventricular excitation and contraction?

a. P wave

b. QRS wave

c. T wave

d. XYZ wave

e. Q wave

Answer: b

12. The heart

a. will contract as a result of stimuli from the sinoatrial node.

b. contracts only as a result of nerve stimulation from the central nervous system.

c. pulse is primarily under the control of the atrioventricular node.

d. is completely independent of all nervous control.

Answer: a

13. The main function of valves within the circulatory system is to

a. permit blood to circulate rapidly.

b. prevent blood from moving too rapidly.

c. act as filters.

d. stop circulation whenever necessary.
e. prevent blood from moving in the wrong direction.

Answer: e

14. Which of the following is most responsible for the movement of blood to the lungs?

a. left ventricle
b. left atrium
c. right atrium
d. right ventricle

Answer: d

True/False *(1 point each)*

15. White blood cells are one of the components of the formed elements in blood.

Answer: T

16. Atria have a much thicker wall when compared to the ventricles.

Answer: F

17. Movement of blood in the veins is due to skeletal muscle contraction.

Answer: T

18. The AV node is known as the pacemaker.

Answer: F

19. The possibility of hemolytic disease of the newborn exists when the mother is Rh^- and the father is Rh^+.

Answer: T

20. Hypertension is an accumulation of soft masses of fatty materials beneath the linings of arteries.

Answer: T

***San Jose Mercury News* Bonus Question** *(1 point. total):*

A protein was recently discovered in pregnant women that has AIDS-fighting, cancer-fighting, and anemia-fighting properties; give the name and source of this protein.

Answer: hCg-associated factor (HAF), which is found in urine

WEEK 12 EXAM: CNS AND DRUGS

1. The two hemispheres of the brain share information across the
a. cerebellum.
b. corpus callosum.
c. hypothalamus.
d. neural ganglia.
Answer: b

2. Heroine affects the concentration of which neurotransmitter?
a. acetylcholine
b. serotonin
c. endorphins
d. norepinephrine
Answer: c

3. Damage to nasal tissues and perforation of the nasal septum are characteristic of abusing which of the following drugs:
a. barbiturates
b. cocaine
c. LSD
d. marijuana
e. amphetamines
Answer: b

4. According to Dr. Anand, what percentage of adults over the age of 65 are afflicted with Alzheimer's disease?
a. 6 percent
b. 11 percent
c. 20 percent
d. 24 percent
e. 30 percent
Answer: b
5. Cocaine affects the concentration of which neurotransmitter?
a. acetylcholine
b. serotonin
c. endorphins
d. norepinephrine
e. dopamine
Answer: e

6. The ____________ is the part of the brain that allows us to be conscious.

a. thalamus
b. pons
c. cerebellum
d. medulla
e. cerebrum

Answer: e

7. Information on the right side of our visual world is processed by which side(s) of the brain?

a. right
b. left
c. both right and left
d. neither right nor left

Answer: b

8. Marijuana affects the concentration of which neurotransmitter?

a. acetylcholine
b. serotonin
c. endorphins
d. norepinephrine
e. dopamine

Answer: b

9. Which of the following are characteristic of Alzheimer's disease?

a. neurofibrillary tangles
b. chemical deficits
c. deteriorated brain tissue
d. a and c only
e. a, b, and c

Answer: e

For questions 10-14, pick one of the following:

a. responsible for vision
b. responsible for hearing and smelling
c. controls movement of voluntary skeletal muscles
d. responsible for elaboration of conscious thought
e. responsible for sensations of temperature, touch, pressure, and pain from skin

10. temporal lobe

11. frontal lobe association areas

12. parietal lobe

13. occipital lobe

14. frontal lobe motor areas

Answers:

10. b
11. d
12. e
13. a
14. c

For questions 15-19, pick one of the following:

a. hypothalamus
b. limbic system
c. cerebellum
d. pons
e. medulla oblongata

15. regulates hunger, sleep, thirst, body temperature, water balance, and blood pressure

16. contains reflex centers for vomiting, coughing, sneezing, hiccoughing, and swallowing

17. controls balance and complex muscular movement

18. involved in emotions, memory, and learning

19. contains reflex centers concerned with head movements in response to visual and auditory stimuli

Answers:

15. a
16. e
17. c
18. b
19. d

True/False

20. The electrocardiogram (EKG) is used to monitor brain waves.
Answer: F

21. Fetal alcohol syndrome occurs because alcohol crosses the placenta freely.
Answer: T

22. Learning is accomplished by a decrease in the number of synapses in the brain.
Answer: F

23. In a grand mal epileptic seizure, the cerebrum is receiving no electrical impulses.
Answer: F

24. According to Mader, dreaming usually occurs in REM sleep.
Answer: T

Short Answer *(6 points)*

For each of the following blood alcohol concentrations, describe the effects on the brain and the nervous system:

.05
.10
.15
.30

Answers:
.05 reasoning and judgment are impaired
.10 speech, vision, coordination impaired
.15 control of large muscles fails, leading to staggering and weaving
.30 stupor, unconsciousness

Concise, specific answers are called for.

What is the State of California's blood alcohol concentration limit for drivers over 21?
Answer: .08

Bonus Question

According to Dr. Anand, our brain has how many neurons?
Answer: 10^{11}

WEEK 13 EXAM: AGING

1. Changes in tissues that result in stiffening and loss of elasticity are due to a condition called
   a. collagen binding.
   b. cross linking.
   c. ligamentosis.
   d. atherosclerosis
Answer: b

2. Progressive decline in bone density is called
   a. osteoblastosis.
   b. osteoarthritis.
   c. osteoporosis.
   d. osteoclastitis.
Answer: c

3. The study of aging is referred to as
   a. geriatrics.
   b. geritol.
   c. gerontology.
   d. senescence.
Answer: c

Remember that answering multiple-choice questions involved choosing the *best* response. Carefully consider the difference between *a* and *c*.

4. Which of the following increases in the adult with advancing age?
   a. elasticity of blood vessels
   b. hearing discrimination
   c. visual sensitivity to light
   d. fat content of the body
   e. length of the spinal column
Answer: d

5. Which of the following does not describe a typical facial change associated with old age?
a. changes in the contours of the face
b. increase in the size of the ear lobe
c. decrease in the size of the skull
d. increase in the thickness of the nasal tip
e. decrease in the thickness of the upper lip

Answer: c

6. You're over at Gorden Biersch having a few beers with your friends. How can you tell if you're sober enough to drive?
a. You can still say "How much wood could a woodchuck chuck if a woodchuck could chuck wood?"
b. You can sing "How much wood could a woodchuck chuck if a woodchuck could chuck wood?" while doing the mambo on the bar counter.
c. Hard to say, since after a few beers, your judgment is impaired.
d. The Highway Patrol tells you to have a nice evening as you're leaving the bar.

Answer: c

7. After about the age of 20,
a. fluid intelligence increases, and crystallized intelligence decreases.
b. fluid intelligence decreases, and crystallized intelligence increases.
c. fluid intelligence increases, and crystallized intelligence increases.
d. fluid intelligence decreases, and crystallized intelligence decreases.

Answer: b

For questions 8-11, use the following:

a. young adulthood
b. senescence
c. infancy
d. adolescence
e. childhood

8. Sex organs become functional.

9. Growth and body proportions change.

10. Physical peak in muscle strength, reaction time, and sensory perception.

11. Greatest period of growth and sensorimotor development.

Answers:

8. d
9. e
10. a
11. c

For each state or condition listed in 12-17, indicate whether it is more common in males or females.

a. males
b. females

12. Have a marked increase in heart disease in their 40s.

13. Tend to live longer.

14. Experience a growth spurt later in adolescence.

15. More likely to have a stroke.

16. Have sex hormones that provide protection against cardiovascular disease.

17. Gamete production occurs up until death.

Answers:

12. a
13. b
14. a
15. b
16. b
17. a

True/False

18. Antioxidant enzymes play an important role in detoxifying free radicals.
Answer: T

19. As the skin ages, the number of melanocytes decrease and there are more sebaceous glands.
Answer: F

20. The ages of both the mother *and* the father at the time of conception are highly significant factors in the life expectancy of their offspring.
Answer: F

21. Type II diabetes is more common in older individuals.
Answer: T

22. The thymus gland becomes very large in older individuals.
Answer: F

23. A woman's uterus becomes smaller with age.
Answer: T

24. By age nine, a child's brain is almost full size.
Answer: T

25. Accidents are the leading cause of death among the elderly.
Answer: F

26. The weight of internal organs decreases with age.
Answer: T

27. Damaged brain cells repair themselves.
Answer: F

28. "Multi-infarct" dementias refer to mental impairments that can be corrected if properly treated.
Answer: F

29. Rate of tissue repair slows down with advancing age.
Answer: T

30. "Life span" and "life expectancy" mean the same thing.
Answer: F

WEEK 14 EXAM: HUMAN ECOLOGY

Multiple Choice: *(Select the one best answer for the following.) (1 point each)*

1. ___________ is the study of how organisms interact with one another as well as with their physical and chemical environment.
a. Anthropology
b. Sociology
c. Geology
d. Ecology
e. Scatology
Answer: d

2. For a given species, the maximum growth rate under ideal conditions is the
a. biotic potential.
b. limiting factor.
c. carrying capacity.
d. density control.
Answer: a

3. Of the following, which is found at the bottom of the food pyramid?
a. carnivores
b. decomposers
c. herbivores
d. producers
e. omnivores
Answer: d

4. The maximum population size that the environment can support for an indefinite period of time is the
a. biotic potential.
b. environmental resistance.
c. carrying capacity.
d. density control.
Answer: c

5. Which of the following factors does NOT affect sustainable population size?

a. predation
b. competition
c. available resource
d. pollution
e. All of them can affect population size.

Answer: a

6. Energy from fossil fuels is __________; their extraction and use come at __________ cost to the environment.

a. renewable; low
b. renewable; high
c. nonrenewable; low
d. nonrenewable; high

Answer: d

7. An organism that uses radiant energy in photosynthesis is known as a

a. carnivore.
b. decomposer.
c. herbivore.
d. producer.
e. omnivore.

Answer: d

8. Ozone (O_3)

a. causes respiratory distress in humans.
b. damages plant leaves.
c. reduces cancer by blocking ultraviolet radiation.
d. a, b, and c are correct.
e. None of them is correct.

Answer: d

9. Since the mid eighteenth century, human population growth has been

a. leveling off.
b. accelerating.
c. growing slowly.
d. not much to speak of.

Answer: b

10. Pollutants (e.g., carbon oxides, nitrates, and sulfur and nitrous oxides) disrupt ecosystems because
a. their components differ from those of natural substances.
b. only humans have a use for them.
c. there are no evolved mechanisms to deal with them.
d. their only effect is on ecosystems, not humans.

Answer: c

11. Acid rain is caused by rain water reacting with
a. sulfur oxides.
b. nitrogen oxides.
c. hydrogen oxides.
d. a and b only.
e. a, b, and c.

Answer: d

12. The greenhouse effect is caused mainly by the buildup of __________ in the atmosphere.
a. carbon dioxide
b. hydrogen
c. nitrogen
d. oxygen
e. water

Answer: a

13. Which of the following factors of the forest environment is abiotic?
a. rabbits
b. trees
c. ferns
d. the soil
e. woodpeckers

Answer: d

14. Actions that can improve conditions in our biosphere include
a. landscaping for low water use.
b. having more babies.
c. recycling.
d. a and c only.

e. b and c only.

Answer: d

True/False

15. If the number of herbivores in an environment decrease, the number of carnivores feeding on the herbivores will increase.

Answer: F

16. The main cause of global warming is the release of CFCs into the atmosphere.

Answer: F

17. When organisms die and decay, chemical elements are made available to the producers once again.

Answer: T

18. Conventional farming has no deleterious effects on the environment.

Answer: F

19. The realm on earth where life exists is known as the hydrosphere.

Answer: F

20. Acid rain causes aluminum and iron to become concentrated in lakes.

Answer: T

21. Energy is lost as heat at each step of the food chain.

Answer: T

22. The fate of spaceship Earth rests in our hands.

Answer: T

Answer the following questions based on the Mader essay "Preservation of the Everglades" (8 points total).

A. Where in Florida are the Everglades (be specific!)? (1 point)

Answer: Southern Florida, from Lake Okeechobee down to Florida Bay

B. Name at least three plants or (nonhuman) animals that inhabit the Everglades (3 points).

Answer:

herons
egrets
alligators

Plants that might have been named include saw grass prairie, red mangrove, and cypress. Other animals include roseate spoonbill, and anhinga. An answer like "shrimps and crabs" is acceptable, but I'd prefer that students were more specific.

C. The article discusses a major modification that humans have made to the Everglades environment. What exactly was this modification? (1 point)

Answer: The land was drained just south of Lake Okeechobee to grow crops.

D. How has this modification affected the wildlife in the Everglades (in particular, the birds)? (1 point)

Answer: It has affected the reproduction pattern of the birds.

E. What is the "solution" proposed by Mader to preventing the destruction of the Everglades (as well as other environmentally sensitive areas)? (2 points)

Answer: To work with nature by growing plants that are native to the area. In the case of the Everglades, this means planting aquatic plants, rather than sugar cane (a conventional, nonnative plant).

CHAPTER 26

UNIVERSITY OF SCRANTON

BIOLOGY 141 AND 142: GENERAL BIOLOGY I AND II

Janice Voltzow, Associate Professor

BIOLOGY 141 AND 142, GENERAL BIOLOGY I AND II, IS A TWO-SEMESTER SURVEY OF biology. Three to four hour-long "midterms" and a final are given each semester. A midterm and final from the second semester are included here. Each midterm counts for 100 points and the final 200 points out of a total of 650 points for the course. Thus the exams determine most of the course grade.

The main objectives of the course are to introduce students to some of the fundamental concepts of biology; to help them learn to read, think, and write critically; and to help them learn how to learn independently. I expect students to obtain a basic knowledge of some of the major concepts of biology, to begin to relate what happens at one level of organization to another (for example, to relate cellular respiration to the need for gas exchange systems), to think critically about what they read in a textbook or outside reading, and to begin to learn to organize for themselves the large amount of information they are exposed to over the course of two semesters.

Before the exam, review the topic outlines provided by the professor and use them as a framework to study the material. Use the figures in the book as ways of testing yourself. Cover the legend and see if you can write a paragraph about the figure. Or cover the figure and see if you can draw it yourself, understanding your drawing and not just memorizing it.

MIDTERM EXAM

For multiple-choice questions, circle the letter next to the best answer on this test and fill in your answer on the bubble sheet provided. For the short-answer and essay questions, write your answer directly on this test.

Multiple Choice *(2 points each)*

The multiple-choice section counts for 50 percent of the grade. Multiple-choice questions generally require students to process specific details that have been covered in the course.

1. The exoskeletal cuticle of an arthropod

a. is composed of chitin.
b. grows with the animal.
c. consists of a single layer.
d. is not very strong or tough.

Answer: a

2. Movement by animals and of materials within animals can be accomplished by

a. amoeboid movement.
b. ciliary beating.
c. contraction of muscle fibers.
d. flagellar beating.
e. all of the above

Answer: e

3. When striated muscle contracts

a. the I bands widen.
b. the Z lines move closer together.
c. the Z lines move farther apart.
d. the distance between Z lines does not change.
e. none of the above

Answer: b

4. Endoskeletons are the principal means of support in

a. snails.
b. crabs.
c. sponges.
d. corals.

e. insects.

Answer: c

This is a question many students found difficult. It may be hard to think of sponges as having a skeleton.

5. The walls of capillaries

a. are lined by a thick layer of smooth muscle.

b. are permeable to substances such as oxygen, glucose, and amino acids.

c. have no spaces between cells.

d. are reinforced with elastic connective tissue.

e. all of the above

Answer: b

6. Lymph nodes

a. lie at points where small lymphatic vessels converge.

b. are sites for the production of lymphocytes.

c. contain cells that phagocytize bacteria.

d. tend to become inflamed when you have an infection.

e. all of the above.

Answer: e

7. Specialized tubules in insects that make direct contact with the tissues to provide oxygen and remove carbon dioxide are called

a. dermal branchiae.

b. tracheae.

c. gills.

d. lungs.

Answer: b

8. Helper T cells

a. play a role in both cell-mediated and humoral immunity.

b. release antibodies into the bloodstream.

c. phagocytize large invaders such as parasites.

d. only participate in cell-mediated immunity.

Answer: a

A question many students found difficult, perhaps because all of the alternatives relate to the immune system.

9. Hemocyanin

a. is always found in blood cells.
b. contains copper, which gives its solution a blue color.
c. only occurs in arthropods.
d. is a small molecule.
e. all of the above

Answer: b

10. Vertebrate skeletal muscle cells

a. are syncytial.
b. can actively contract and reextend themselves.
c. are the fastest muscle cells in the animal kingdom.
d. show little internal organization.

Answer: a

11. Which of the following statements is true about the immune response in animals?

a. At the biochemical and cellular level, only vertebrates appear to be able to distinguish self from nonself.
b. Nonspecific immune defenses include amoeboid scavengers such as amoebocytes or phagocytes.
c. Only humans have evolved the ability to produce a cell-mediated immune response.
d. Internal defense only provides protection against microorganisms.

Answer: b

Most of these alternatives can be eliminated. This is a good way to approach a multiple-choice question if you don't immediately know the answer.

12. The molecule that forms the principal component of most connective tissue and that gives it many of its important properties is

a. cellulose.
b. chitin.
c. collagen.
d. actin.
e. tubulin.

Answer: c

13. All skeletons must resist

a. compression.
b. tension.
c. shear.
d. torsion.
e. all of the above.

Answer: e

A question calling for fundamental understanding of the function of the skeleton.

14. Immunological memory

a. can only result from a humoral immune response.
b. is the mechanism that explains why children are vaccinated against diseases.
c. usually lasts only a short time (several months).
d. involves only B cells.

Answer: b

15. Blood moves more slowly through capillaries than through arteries because

a. the total cross-sectional area of the capillaries is less than that of the arteries.
b. the total cross-sectional area of the capillaries is greater than that of the arteries.
c. capillaries are thinner than arteries.
d. capillaries have thinner walls than do arteries.

Answer: b

A tricky question that many students found difficult. Read the question carefully. By definition, capillaries are small; therefore, it is all too easy to assume that *a*, *c*, or *d* is a correct response. While *c* and *d* are accurate statements, they do not explain why blood moves more slowly through capillaries. At first glance, answer *a* seems plausible, but, in fact, the explanation for the reduced velocity of blood flow in capillaries is *b*. While individually smaller than arteries, the capillary net is much larger.

16. The type of joint that permits the bones of the forearm to rotate when the wrist twists is a

a. ball and socket.
b. suture.
c. pivot.
d. saddle.

e. hinge.

Answer: c

Students found this difficult.

17. Each muscle cell is also called a

a. myofibril.
b. motor unit.
c. sarcomere.
d. myofilament.
e. muscle fiber.

Answer: e

It is important to an understanding of muscle function to recognize that a muscle *cell* and a muscle *fiber* are one and the same.

18. Which of the following is true of humoral immunity?

a. Plasma cells release antibodies into the bloodstream that attack foreign antigens.
b. Plasma cells release antigens into the bloodstream that attack foreign antibodies.
c. It involves no type of T cell, only B cells.
d. B cells divide and give rise to clones of T cells.

Answer: a

Read carefully! Answers *a* and b differ by only a single critical—and similar-looking—word.

19. One of the main differences between a closed circulatory system and an open one is

a. open systems only work for primitive animals that aren't very efficient.
b. open circulatory systems lack arterioles and venules.
c. closed systems always contain hemoglobin; open systems always contain hemocyanin.
d. the site of exchange in a closed system is the capillary; the site of exchange in an open system is unlined spaces called sinuses.
e. open systems lack a heart.

Answer: d

20. The Principle of Continuity

a. states that the rate at which a fluid enters one end of a pipe must equal the rate at which the fluid leaves the other end.

b. explains the differences in rates of flow observed in the different parts of the circulatory system.

c. can be applied to any system of parallel pipes or tubes.

d. applies to both open and closed circulatory systems.

e. all of the above

Answer: e

21. Which of the following is true of cell-mediated immune response?

a. Plasma cells release antibodies into the bloodstream that attack foreign antigens.

b. It involves no type of T cell, only B cells.

c. It is the primary defense against viruses that have entered cells.

d. No memory cells are produced because no antibodies are produced.

Answer: c

22. The movement of oxygen and carbon dioxide in the body is accomplished by

a. exocytosis and endocytosis.

b. bulk flow.

c. osmosis.

d. diffusion.

e. facilitated diffusion.

Answer: d

23. The group of animals with the most efficient respiratory system is the

a. amphibians.

b. birds.

c. mammals.

d. reptiles.

e. fish.

Answer: b

24. Actual exchange of gases in the lungs of mammals occurs in the

a. bronchi.

b. bronchioles.

c. tracheas.

d. alveoli.

e. air capillaries.

Answer: d

25. An example of a homeothermic ectotherm is a

a. mouse.
b. snail.
c. bear.
d. lizard.

Answer: d

26. The worst joke I've heard so far is the one about the

a. hoop stress.
b. blind capillaries.
c. humoral response.
d. Huh? I don't care as long as I can have the two points.

Answer: Any answer earns 2 points!

Short Answers

Represents 25 percent of the grade. Short-answer questions and problems require students to apply basic concepts or organize them in new ways.

27. (7 points) Below is a diagram of a relaxed skeletal muscle. Draw a similar diagram of the same muscle in the contracted state to the same scale directly beneath it. Label a sarcomere, myosin, actin, A band, and I band.

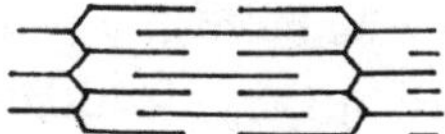

Answer:

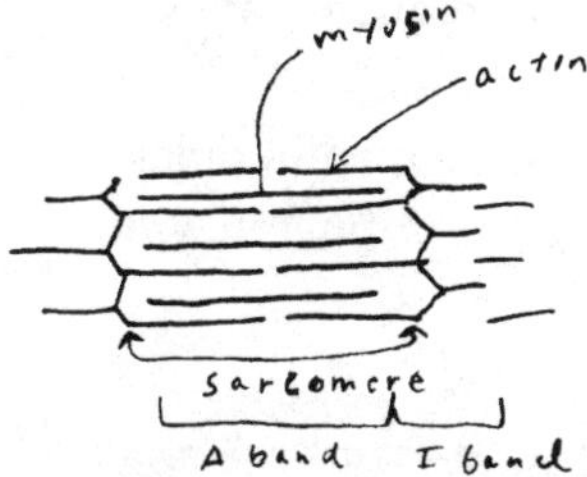

It is important that the sarcomere size is shorter, but the molecule length is the same. This diagram is directly from a handout distributed in class.

28. (5 points)

A. Explain why the line in the diagram is S-shaped instead of straight.
B. Draw a line indicating the Bohr effect.
C. What is the Bohr effect?

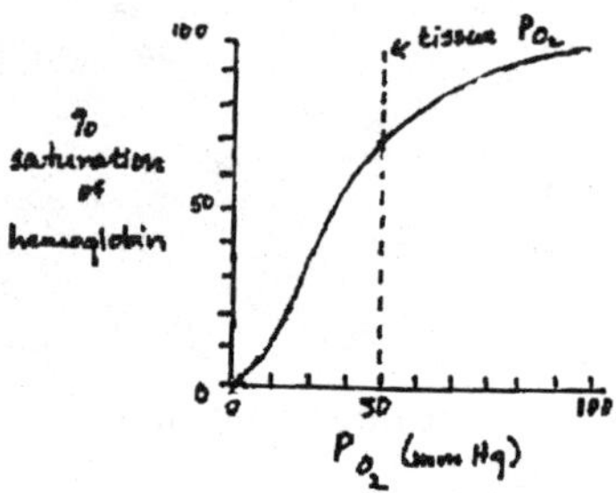

Answers:

A. The S shape is due to the cooperativity of the Hb molecule.

B.

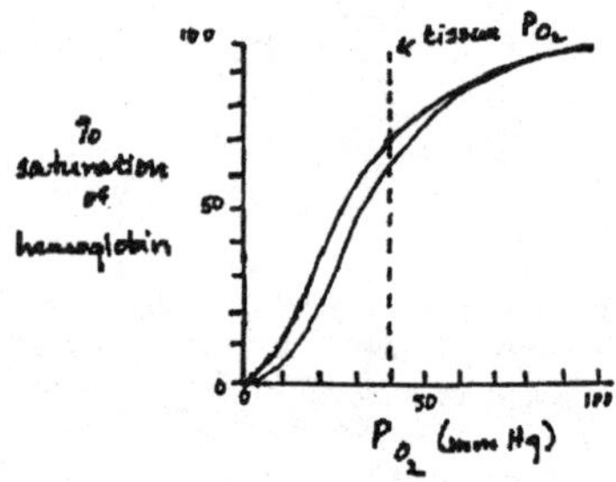

The curve drawn must be to the right of the original curve and must fuse with it at 100 percent saturation.

C. The Bohr effect is created by the acid sensitivity of Hb. It has a reduced affinity for O_2 at lower pH, which is caused by CO_2 forming carbonic acid.

29. **(4 points) Briefly describe the sequence of events that leads to blood clot formation in humans.**

Answer: platelets rupture, releasing thromboplastin, which, in the presence of ionized Ca, yields prothrombin → thrombin (enzyme), which polymerizes soluble fibrinogen → insoluble fibrin

The answer can be expressed entirely in words or with arrows indicating the process. The important thing is to map out the sequence of events.

30. (4 points) On the graph, label the line that represents plasma hydrostatic pressure and the line that represents plasma osmotic pressure. Indicate on the graph the region of the capillary in which fluid would move out of the capillary into the interstitial fluid and the region where the fluid would move into the capillaries from the interstitial space.

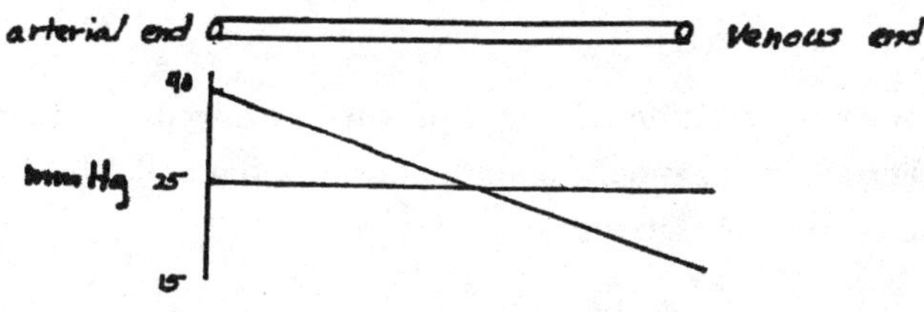

Answer:

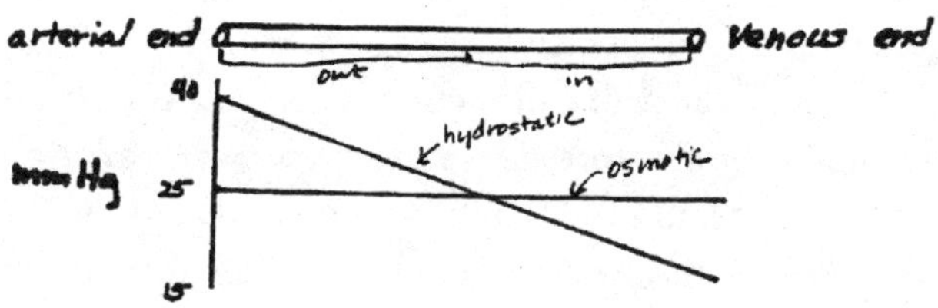

Essay question. *10 points. Answer only one question. Reread your answer when you are finished to be certain it answers the question.*

Represents 25 percent of the grade. Essay questions permit them to write critically about what they have learned or what they think about what they have learned. Examples of two responses are given.

A. Given Fick's Law: $Q = DA\dfrac{(C_1 - C_2)}{L}$

Describe in detail how fish and birds take advantage of this law to increase the efficiency of oxygen diffusion in their gills and lungs.

Answer: Fish maximize Q by means of the large surface area of the gills, which maximizes A. Countercurrent exchange maximizes $C_1 - C_2$, while the fish's very thin gill lamellae minimize L.

Birds maximize Q by means of air capillaries, which maximize area, A. Through cross-current exchange and air sacs, they maximize $C_1 - C_2$, while minimizing L by virtue of thinness of the air capillaries.

In evaluating the response, I try to break down the question into specific parts and assign points accordingly. In the fish portion of the answer, for addressing the issue of maximizing *A*, I award 1½ points; for explaining how countercurrent exchange affects $C_1 - C_2$, 2 points; for the role of very thin gill lamellae in minimizing *L*, 1½ points. Similarly, in the bird portion, 1½ points are earned for identifying the function of air capillaries as maximizing *A*; 2 points for the role of cross-current exchange and air sacs in $C_1 - C_2$; and 1½ points for noting that the very thin air capillaries minimize *L*.

B. Imagine that two different phyla of worms evolved at the same time, one looking like figure A below and the other looking like figure B. Which phylum do you think would persist through time and why?

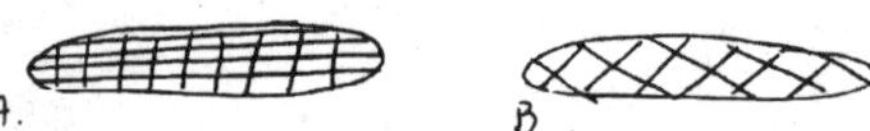

Answer: Phylum B would persist, whereas Phylum A would not. In B, helical fibers control the animal's range of contraction. This structure would resist tension as well as compression. It would store elastic energy, prevent kinking, and resist torsion. All of these properties would tend to contribute to B's survival. The nonhelical structure of A would not endow the animal with these important properties. It would have difficulty moving and would be vulnerable to many kinds of injury, as well as to predators.

Bonus question !!!! *(4 points)*

A dog has one aorta, whose cross-sectional area is about 0.8 cm², and about 1.2 × 10⁹ capillaries, whose total cross-sectional area is about 600 cm². If the blood moves through the aorta at 40 cm/s, what is the velocity at which it moves through an individual capillary?

Answer: $s_1 v_1 = s_2 v_2$

$$s_1 = 0.8\text{cm}^2 \qquad v_1 = 40\text{cm/s}$$

$$s_2 = 600\text{cm}^2$$

$$v_2 = \frac{s_1 v_1}{s_2}$$

$$= \frac{(0.8\text{cm}^2)(40\text{cm/s})}{600\text{cm}^2}$$

$$= 0.053\text{cm/s}$$

FINAL EXAM

Attend all classes. Sleep well and have breakfast before coming to class. Briefly look at the chapter covering the subject to be discussed—read the introduction and look at figures at beginning of

chapter—before the lecture on that subject. Ask questions whenever anything is unclear. As soon as possible after the lecture, review your notes, using the text to fill in details you might have missed or don't understand. Begin asking yourself questions about the material covered.

For multiple-choice questions, circle the letter next to the best answer on this test and fill in your answer on the bubble sheet provided. For the short-answer and essay questions, write your answer directly on this test.

Multiple Choice *(2 points each)*

The multiple-choice section counts for 40 percent of the grade.

1. Homeostasis is the process of controlling

a. the osmotic concentration of body tissues.
b. pH.
c. dissolved substances in the body fluids.
d. all of the above

Answer: e

2. A nerve impulse is propagated because

a. calcium rapidly exits from the neuron.
b. refractory period ends.
c. neuron's membrane has a change in permeability to sodium.
d. potassium accumulates outside the cell.

Answer: c

Many students found this difficult.

3. The rumen and reticulum of ruminants

a. contain huge numbers of bacteria and protozoans.
b. produce the cud that the animals ruminate.
c. are specializations of the digestive system that aid in the digestion of cellulose.
d. are not the true stomach of the ruminant.
e. all of the above

Answer: e

4. Which of the following has an incomplete digestive system?

a. annelids
b. flatworms
c. molluscs
d. arthropods
e. ruminants

Answer: b

5. Which of the following is not a condition for Hardy-Weinberg equilibrium?

a. small population
b. isolation from other populations
c. no net mutations
d. no natural selection

Answer: a

6. Microevolution

a. is the mechanism by which microorganisms evolve.
b. is the process of evolution below the species level.
c. almost never happens.
d. acts at the level of specific genes.

Answer: b

7. Neutral mutations

a. are not subject to selection.
b. only occur in simple organisms.
c. confer a disadvantage.
d. do not occur; either a gene enhances survival or it does not.

Answer: a

8. Macroevolution refers to

a. the process by which large organisms evolve.
b. large-scale patterns of diversity through evolutionary time.
c. allopatric speciation.
d. fossils only.

Answer: b

9. The exoskeleton of an arthropod

a. is composed of chitin.
b. grows as the animal grows.

c. is very permeable to water.
d. is stiff but not very strong.

Answer: a

10. The endoskeleton of an echinoderm

a. is composed of chitin.
b. grows as the animal grows.
c. consists completely of inorganic material.
d. is stiff but not very strong.

Answer: b

Students found this difficult.

11. When a striated muscle contracts

a. myosin filaments shorten.
b. actin filaments shorten.
c. the A band gets smaller.
d. the sarcomere gets shorter.

Answer: d

12. The type of nutrient that yields the most kilocalories per gram metabolized is

a. carbohydrates.
b. lipids.
c. proteins.
d. vitamins.

Answer: b

13. Specialized tubules in insects that make direct contact with the tissues to provide oxygen and remove carbon dioxide are called

a. dermal branchiae.
b. tracheae.
c. gills.
d. lungs.
e. book lungs.

Answer: b

14. During inhalation in humans

a. the pressure in the thoracic (chest) cavity is greater than the pressure inside the lungs.

b. the pressure in the thoracic cavity is less than the pressure within the lungs.
c. the diaphragm moves upward and becomes more curved.
d. muscles in the nostrils suck the air into the lungs.
e. none of the above

Answer: b

A number of students had difficulty with the mechanics of inhalation in humans.

15. Which involves a positive feedback stimulation?

a. temperature control
b. sexual stimulation
c. presence of estrogen on the anterior pituitary during most of the female reproductive cycle
d. clotting of blood

Answer: b

Read the question carefully. It often contains hints as to the correct response. First: which of the choices involve feedback? Then: which involve *positive* feedback?

16. Uric acid

a. is excreted by mammals.
b. is less energetically expensive to synthesize than urea.
c. is toxic if it accumulates in tissues.
d. is produced by reptiles and birds.
e. requires large amounts of water to excrete.

Answer: d

17. Marine bony fish

a. have blood that is more concentrated than seawater.
b. secrete excess salt by active transport across the gills.
c. produce urine more concentrated than their blood.
d. never need to drink because they are surrounded by water.

Answer: b

18. A blood clot is formed by a meshwork of

a. prothrombin.
b. thrombin.
c. fibrin.

d. albumin.
e. fibrinogen.

Answer: c

It is easier to make a correct response if you have a grasp of the *sequence* of the clotting process.

19. Small islands that serve as the nesting sites of marine birds are often caked with guano, a smelly white paste. One of the principal components of guano is

a. urea.
b. uric acid.
c. ammonia.
d. toothpaste.

Answer: b

Choose your answer with care. Note that the first three choices are related—but uric acid most accurately describes the principal compound constituent of guano.

20. The capillary bed that is the site of ultrafiltration in a vertebrate nephron is called the

a. proximal convoluted tubule.
b. Bowman's capsule.
c. nephron.
d. glomerulus.
e. ureter.

Answer: d

21. The single long process that extends from a typical motor neuron is the

a. axon.
b. synapse.
c. dendrite.
d. cell body.

Answer: a

22. Functionally speaking, a nerve impulse is

a. a flow of electrons along the outside of the plasma membrane of a neuron.
b. the movement of cytoplasmic elements through the core of the neuron.
c. a series of changes in membrane potentials.

d. a lengthening and shortening of the membrane extension of an individual neuron.

Answer: c

23. **The acrosome reaction**
   a. permits a sperm to penetrate the jelly coat of an egg.
   b. promotes species recognition via surface molecules and receptors on the sperm and egg.
   c. induces the blocks to polyspermy by the egg.
   d. all of the above

Answer: d

24. **All of the following are essential organic nutrients *except***
   a. carbohydrates.
   b. lipids.
   c. proteins.
   d. vitamins.
   e. chocolate.

Answer: e

Short Answers

Short answers account for 30 percent of the grade.

25. **(4 points) A population of pink and white flowers follows the patterns of classic Mendelian genetics in the genes for flower color. Pink is the dominant phenotype. If 96 percent of the population is pink, what is the frequency of the recessive allele in the population?**

Answer:

$p_2 + 2pq + q_2 = 1$ $\qquad$ $p + q = 1$

$p_2 + 2pq = 0.96$

$q_2 = 0.04$

$q = 0.2$

26. **(10 points) List five potential mechanisms or evolutionary agents by which microevolution can occur.**

Answer:
1. genetic drift
2. mutation
3. migration (gene flow)
4. natural selection
5. sexual selection (nonrandom mating)

Students could notice that these are the inverse of the answer for question 5. This relationship was emphasized in class.

27. (4 points) Members of the phylum Platyhelminthes lack a system for internal transport. How do they accomplish gas exchange and distribution of nutrients?

Answer: Gas exchange is accomplished through diffusion, and nutrients are distributed through a branched gut.

This question requires synthesis of information gathered from several textbook chapters and from lab.

28. (6 points) The three main functions of the vertebrate nephron are ultrafiltration, reabsorption, and concentration of urine. Name where in the nephron each of these functions takes place.

ultrafiltration
Answer: Bowman's capsule

reabsorption
Answer: proximate and distal convolution tubules

concentration of urine
Answer: loop of Henle and collecting duct

29. (4 points) A study of the osmotic concentration of the blood of the estuarine crab *Eriocheir sp.* yielded the following curve. Are these animals *osmoregulators* or *osmoconformers*? Explain your answer.

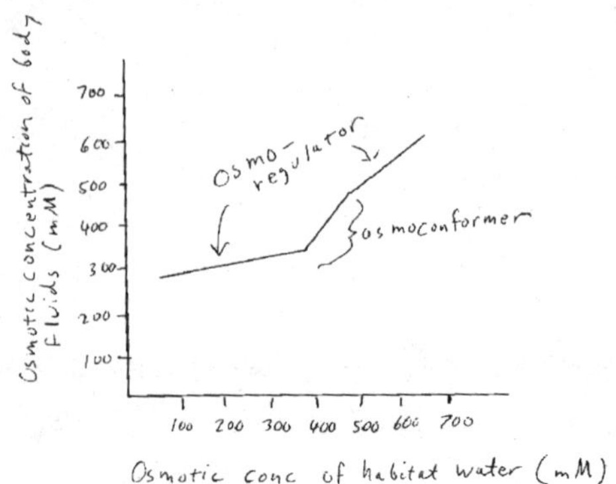

Answer: These animals are both osmoregulators and osmoconformers. They behave as a combination of the two, depending on environmental conditions.

A successful response requires awareness that organisms don't always fall neatly into textbook categories.

30. **(10 points) In the space below, draw a cross-section through an earthworm. Be certain to include and label the following structures: gut, nerve cord, blood vessels, septum, body wall, metanephridium, setae, coelom.**

Answer:

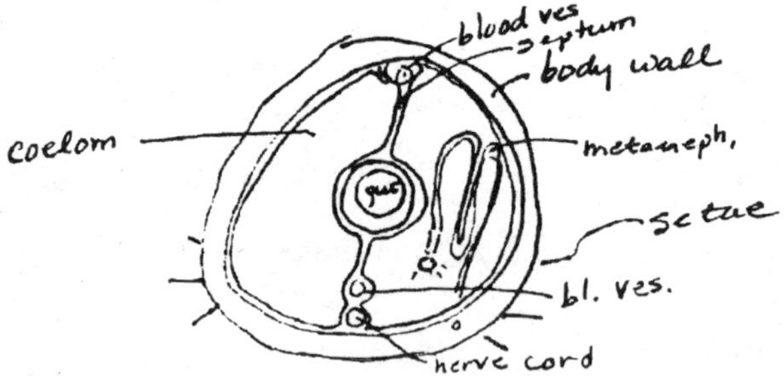

Matching *(2 points each)*

This section counts for 10 percent of the grade.

Write the correct term from the list below in the space in front of its definition. (*Note:* There are more terms than definitions; you will not use all terms. No term is used more than once.)

adaptation
Bohr effect
countercurrent exchange
cell-mediated immunity
coelum
deuterostome
dioecious
evolution
Fick's law
hemolymph
humoral immunity
microvillus
monoecious
natural selection
notochord
phylogeny
principle of continuity
protostome
pseudopodium
threshold potential

An easy page for students to gain confidence and refresh themselves. Note that most terms also have a partner that would be a wrong response.

Change in gene frequency over time within a gene pool.
Answer: evolution

Animal with spiral cleavage in which mouth usually arises at site of blastopore.
Answer: protosome

Animals that are hermaphroditic.
Answer: monoecious

Law that explains the geometry of circulatory systems.
Answer: principle of continuity

Defensive response in which antibodies are released into the circulatory system.
Answer: humoral immunity

A cytoplasmic projection of the cell used to carry out amoeboid locomotion.
Answer: pseudopodium

A cytoplasmic projection of the cell that serves to increase surface area for absorption.
Answer: microvillus

Minimum stimulus required for an action potential to occur.
Answer: threshold potential

Law that describes the rate of diffusion across a membrane.
Answer: Fick's law

Mechanisms that enhance diffusion across a gill.
Answer: countercurrent exchange

Acid sensitivity of oxyhemoglobin that increases the unloading of oxygen to the tissues.
Answer: Bohr effect

Blood of an animal with an open circulatory system.
Answer: hemolymph

Fluid in the body cavity of an earthworm.
Answer: coelum

Essay question *(10 points) Answer only one question. Reread your answer when you are finished to be certain it answers the question.*

Represents 20 percent of the grade. One example answer is presented here.

A. Discuss the proposition that complex animals, such as humans, are better adapted than simpler ones, such as jellyfish.

B. Name the three phenomena that are necessary and sufficient for natural selection to occur. What is natural selection? At what level of organization does it act? What is the difference between natural selection and evolution? Can natural selection and evolution be observed in the lab? In nature?

Answer:

B. The three phenomena necessary and sufficient for natural selection to occur are variable population, the heritability of at least some variations, and the production of more offspring than can survive.

Natural selection is differential survival and reproduction. It operates on individual organisms, and it is just one of the mechanisms through which evolution proceeds. While natural selection and evolution are processes that can consume many centuries, it is possible to observe the process of evolution in the laboratory, using organisms that reproduce in significant quantity and have relatively brief life cycles. It is also possible to observe the effects of evolution in nature, especially in isolated populations, as Charles Darwin did.

Note how the answer uses the terms of the question to provide structure. It is important to address all areas identified in the question—and to make it clear that you are indeed addressing all of these areas.

C. This spring our front lawn and driveway have been totally covered by maple seeds. It seems that this is a mast year for maples. Describe a set of experiments I could do over the next five years to assess the effect of this masting on the squirrel, robin, and tent caterpillar populations in my yard. Devise three hypotheses and explain how you would test them.

D. Describe and compare the structure and function of one system from column A in one organism from column B and one organism from column C.

A	B	C
support and locomotion	gastropod	cow
respiration	grasshopper	frog
circulation	flatworm	bird
digestion	earthworm	fish
excretion	sea urchin	

CHAPTER 27

TRANSYLVANIA UNIVERSITY

BIOLOGY 1014 AND 1024: GENERAL BIOLOGY I AND II

Peggy Shadduck Palombi, Assistant Professor

THIS IS A TWO-SEMESTER SEQUENCE OF GENERAL BIOLOGY DESIGNED FOR majors as well as nonmajors. General Biology I usually includes four hourly exams, one of which can be dropped, for a total of 50 percent of the grade, and a cumulative final worth 20 percent of the grade. The remainder of the grade comes from lab work and papers. General Biology II usually includes three hourly exams worth 40 percent of the grade and a cumulative final worth 20 percent of the grade. The remainder again comes from lab exams and papers. Each of the three professors teaching a section of this course each semester writes his or her own syllabus and exams, although we cover roughly the same material.

Examples of two finals, one for General Biology I and one for General Biology II, are included here.

My primary objectives in these courses are to help students learn about the essential elements of life and their interrelationships. We cover ecology, population and community dynamics, and relationships between living things and their environment first. We then consider the basic chemistry of living organisms and the organization of cells, followed by coverage of some of the major activities of cells: energy gathering, energy usage, and reproduction. We then focus on the molecular basis of cellular activities, DNA. Evolution is a recurring theme throughout the course. During the second semester, we introduce students to the rich array of life on earth and the primary physiological systems in those organisms. We cover biological classification and the history of life on Earth, moving on to an introduction to the different phyla, with emphasis on animals.

The goal of the courses is not to make students experts in any one area of biology, but to help them learn to read, think, and communicate effectively while at the same time introducing them to some of the many exciting areas of the study of life. We hope students learn to formulate a hypothesis, to design a simple experiment, to question what they read, to read a review article, and to write a simple paper in a scientific style.

FINAL EXAM FOR GENERAL BIOLOGY I

My exams include three parts. The first part (usually) consists of multiple-choice questions. These questions require students to have a thorough understanding of what they read in the textbook and what was presented during lecture. These questions test students' reading and learning of details in the subject. The second part of each exam is a set of short-answer questions. These require that students glean the major concepts from the material presented, and that they express those concepts in a clear and concise manner. The last portion of each exam is an essay question. This requires that students organize both "big-picture" and detailed concepts, often integrating information from multiple lectures and textbook chapters. Sometimes these questions also incorporate experimental design and problem-solving skills that should have been gained during lab exercises.

Study early and often rather than "cramming." Late-night studying does not help most students. Prepare a list of new terms from each lecture and chapter. Write your own outline of each chapter and then compare it with the one given by the authors at the end of the chapter. Answer all the chapter review questions and go ask your professor about any that confuse you. Try to write your own essay questions, then write answers to them.

Do not just read a chapter, but really *think* about it and its implications. Try to apply your knowledge *before* you get to the exam. When answering multiple-choice questions, cover up the answers and try to think of the correct answer before looking at the choices. Read the entire question before answering. Do not give information that is not needed to answer the question. Be concise!

Multiple Choice *(2 points each)*

1. The ultimate source of energy flowing into nearly all ecosystems is
   a. wind.
   b. sunlight.
   c. electricity.
   d. geothermal vents.
   e. radioactivity.

Answer: b

2. Which of the following conceptual levels of ecologic organization incorporates abiotic factors?
   a. ecosystem

b. community
c. population
d. species
e. symbioses

Answer: a

3. The most precise statement of when a human population will achieve zero population growth is when

a. no one has more than two children.
b. couples have an average of about 2.25 children each (to allow for the fact that some people have no children).
c. no one has more than one child.
d. the birth rate equals the intrinsic rate of increase.
e. the birth rate equals the death rate.

Answer: e

4. Several primates have been taught to communicate with humans using sign language. This supports the view that animals other than humans undergo what process?

a. imitation
b. association
c. cognition
d. trial and error
e. classical conditioning

Answer: c

5. Which of the following is NOT a characteristic of chemical reactions?

a. Chemical reactions involve the making and breaking of chemical bonds.
b. Some chemical reactions create electrons; others destroy them.
c. The atoms of the reactants are exactly the same as the atoms of the products.
d. The reactants contain the same number of atoms as the products.
e. Although the atoms of a reaction's reactants and products are identical to each other, their molecular formulas differ.

Answer: b

6. The glucose produced by a plant is used to make the carbohydrate, proteins, fats, and nucleic acids of the plant. Which statement is true about the process of converting sugars into other molecules?

a. Glucose contains all of the elements needed to produce these other molecules.

b. Glucose contains all of the elements needed to produce the carbohydrates, but other elements must come from the soil if the plant is to produce fats, proteins, and nucleic acids.
c. Glucose contains all of the elements needed to produce the carbohydrates and fats, but other elements must come from the soil if the plant is to produce proteins and nucleic acids.
d. Glucose contains all of the elements needed to produce the carbohydrates and nucleic acids, but other elements must come from the soil if the plant is to produce fats and proteins.
e. None of these substances can be produced from glucose alone.

Answer: c

7. A child is hospitalized for a series of chronic bacterial infections and dies despite heroic efforts. At autopsy, the physicians are startled to see that the child's white blood cells are loaded with vacuoles containing intact bacteria. Which of the following explanations is mostly likely to account for this finding?
a. A defect in the Golgi apparatus prevented the cells from processing and excreting the bacteria.
b. A defect in rough endoplasmic reticulum prevented the synthesis of the antibodies (defensive proteins) that would have inactivated the bacteria.
c. A defect in the cell walls of the white blood cells permitted bacteria to enter the cells.
d. A defect in the lysosomes of the white blood cells prevented the cells from destroying engulfed bacteria.
e. A defect in the surface receptors of the white blood cells permitted bacteria to enter the cells.

Answer: d

This is a question about the function of lysosomes. Don't let yourself get distracted by the "real-world" context of the question. If you understand what lysosomes do, you can answer this question correctly.

8. The functioning of an electron transport chain can be analogized with
a. a Slinky going down a flight of stairs.
b. going over a waterfall.
c. climbing a flight of stairs one step at a time.
d. jumping from the top to the bottom of a flight of stairs.
e. playing Ping-Pong.

Answer: a

9. The phase of mitosis during which the nuclear envelope fragments and the nucleoli disappear is called

a. interphase.
b. prophase.
c. metaphase.
d. anaphase.
e. telophase.

Answer: b

10. Mendel's principle of independent assortment states that

a. chromosomes sort independently of each other during mitosis, but not during meiosis.
b. genes sort independently of each other in animals but not in plants.
c. independent sorting of genes produces polyploid plants under some circumstances.
d. each pair of alleles segregates independently during gamete formation.
e. crossing over increases genetic variability.

Answer: d

11. Multiple origins of replication on the DNA molecules of eukaryotic cells serve to

a. remove any errors in replication that occur.
b. create multiple copies of the DNA molecule at the same time.
c. shorten the duration of time necessary for DNA replication.
d. reduce the number of "bubbles" that occur in the DNA molecule during replication.
e. assure the correct orientation of the two strands in the newly growing double helix.

Answer: c

12. The term *genome* means

a. the complete genetic endowment of an organism.
b. a phage designed to transport specific DNA fragments.
c. any collection of genes.
d. the physical appearance of an organism insofar as it is determined by genes.
e. all of an organism's proteins.

Answer: a

13. Most people with recessive genetic disorders are born to parents who are

a. recessive for the disorder.

b. homozygous for the abnormal allele and exhibit the disorder.
c. heterozygous for the abnormal allele but phenotypically normal.
d. pleiotropic for the disorder.
e. trisomic for the abnormal allele.

Answer: c

14. Which of the following types of reproductive barriers separate two species that occasionally mate, but in whom the sperm do not produce the right enzymes to enter the egg?

a. temporal isolation
b. habitat isolation
c. behavioral isolation
d. mechanical isolation
e. gametic isolation

Answer: e

15. The definition of kingdoms is

a. finally complete after many decades of debate and study.
b. only meant for convenience and does not really reflect phylogeny.
c. a work in progress and will always be in flux.
d. all of the above
e. none of the above

Answer: c

16. Which of the following processes is endergonic?

a. the burning of wood
b. the synthesis of glucose from carbon dioxide and water
c. release of heat from the breakdown of glucose
d. the breakdown of glucose to power ATP formation
e. cellular respiration

Answer: b

17. Photosynthetic organisms derive their carbon from

a. carbon monoxide.
b. carbon dioxide.
c. hydrocarbons.
d. methane.
e. all of these, depending on environmental conditions

Answer: b

18. In ecosystems, the flow of _____ is one-way, while _______ are constantly cycled.

a. minerals/energy compounds
b. genetic information/genotypes
c. organic compounds/minerals
d. energy/nutrients
e. food/energy

Answer: d

Circle any and all that apply. (4 points each)

1. Which of the following are properties of life?

a. a precise structural organization
b. the ability to take in energy and use it
c. the ability to communicate verbally
d. the ability to respond to stimuli from the environment
e. the ability to reproduce

Answer: a, b, d, and e

2. Which of the following provides evidence that evolution has led to the development of modern species?

a. the fossil record
b. comparative anatomy
c. biogeography
d. DNA homology
e. comparative embryology

Answer: all answers are correct

Short Answer *(6 points each)*

1. Describe the characteristics of a population pyramid that is wide at the base and narrow at the top.

Answer: A population with this age structure diagram is growing, because it has an abundance of individuals in the prereproductive and reproductive age groups. It is likely that this broad base may be the result of recent improvements in sanitation or medicine, either of which would lead to a decline in infant mortality. The pear shape of the pyramid indicates a very small elderly, or postreproductive, population. This shape pyramid is typical of developing nations, like Mexico.

This is a good answer because it indicates the division of the pyramid into prereproductive, reproductive, and postreproductive years. It indicates that a broad base through age 15 shows low infant mortality and a high birth rate and gives plausible explanations for this shape. It is also good to give an example of a country with such a pyramid.

The writing is fairly clear, and the answer is concise. The answer would be better, however, if it included an introductory sentence explaining the axes of a population pyramid. A successful answer requires the student to understand what a population pyramid shows and how it is organized. The student should consider what factors would result in the bar widths at different ages, including any male/female differences, and describe a population with the appropriate values of those factors.

2. Explain the meaning of the statement "Individuals do not evolve, populations do."

Answer: The statement means that while individuals may change over the course of their lifetime, these changes will not be passed on to offspring. Only genetic changes can be passed on to offspring. Therefore, physical traits that may change—like a weight-lifter's muscle mass—won't be inherited, and therefore won't make an influence on future generations. Populations—rather than individuals—evolve because certain characteristics or traits are more favored by the environment, and so these genes become more and more prevalent in the population's gene pool over time. The entire population, then, becomes more adapted to its surroundings.

This answer gets the major points of heritability only of traits encoded in the genes and the definition of evolution in terms of the percentages of certain alleles in the population gene pool. However, the answer is too wordy and repetitious. The question could easily have been answered in two clear sentences. The answer also lacks a clear expression of the concept that only genetic mutations found in the gametes have the possibility of being passed to offspring, not mutations in somatic cells.

The student used a good strategy for answering the question; she divided the phrase into its two clauses and addressed each clause.

3. Explain how the amino acid sequence of one or more polypeptides gives rise to the three-dimensional structure of a protein.

Answer: The primary structure of a protein consists of the actual sequence of the amino acids. The 20 different amino acids can be arranged in different sequences and lengths. The charges of the R-groups (variable of the amino acid) cause the secondary structure, which is either an alpha helix or a pleated sheet. The tertiary structure is actually the secondary structures combined because of different charges. This forms a globular structure that then attaches to other globular structures to form the quaternary structure of the protein. The structure determines shape and shape determines function.

This answer is well-organized and fairly concise, moving from primary to quaternary structure and indicating that this determines function. The second sentence is not necessary and could be eliminated from the answer.

A better answer would include a brief description of an alpha helix and beta pleated sheet and explain how the charges of the R groups lead to these structures. The explanation of tertiary structure is vague and needs a clause on disulfide bonds. The explanation of quaternary structure is weak; what is meant by "attaches"?

The student clearly had the major concepts, but lacked detailed understanding.

4. Explain the pairing of the four DNA bases in a double-stranded molecule of DNA. How do the strands bond together?

Answer: The four nitrogenous bases in DNA are adenine, thymine, cytosine, and guanine. Adenine always bonds with thymine, and cytosine always bonds with guanine. The strands bond together by hydrogen bonds. They always pair up the same way due to their sizes.

This answer is clear, concise, and well-organized. It answers the question without giving unnecessary details. The weakest part is the last sentence, which could use additional explanation of the size and shape differences between purines and pyrimidines and how these lead to a different number of potential hydrogen bonds, and hence the pairing.

5. Protein synthesis requires the interaction of three classes of RNA molecules. Tell how RNA differs from DNA. Name the three classes of RNA and give a one-sentence explanation of the function of each.

Answer: RNA differs from DNA by its structure and the contents of its nucleotides. RNA is usually a single strand, while DNA is usually a double helix form. RNA has uracil in place of DNA's thymine. The other bases are the same. RNA is made with the sugar ribose, whereas DNA is made with deoxyribose. The three types of RNA are messenger RNA (mRNA), ribosomal RNA (rRNA), and transfer RNA (tRNA). mRNA is used in the production of proteins by being the complementary strand to a DNA sequence from the nucleus. mRNA is carried outside the nucleus into the cytoplasm and translated. tRNA carries amino acids to the ribosome and helps attach them to the polypeptide chain in the right place. rRNA is used in making the ribosomes that serve as the workbench where translation of the mRNA sequence takes place.

The answer is again well-organized and clearly written. The three major differences between RNA and DNA are listed, and the three types of RNA are correctly listed. The explanations of their functions are clearly and correctly presented.

6. In the fluid mosaic model of plasma membranes, what make the membrane fluid? What parts constitute the mosaic?

Answer: A bilayer of phospholipids composes the plasma membrane. These phospholipids have heads, which are on the outsides of the bilayer, and tails that constitute the middle of the bilayer. The membrane is fluid due to the fact that it is like a floating layer surrounding the cell and organelles. It is also fluid because it allows some things to pass through it and into the cell and also out of the cell. The mosaic is made of different things that are between the phospholipids. Proteins are wedged between the phospholipids. The fact that it is not merely phospholipids that make up the plasma membrane gave rise to the term "mosaic."

This student presented the major points, but with some confusion and some poor writing. He organized his answer well by first answering part 1 of the question and then part 2. An explanation of head and tails and why they are oriented as described would help this answer.

The third sentence is unclear and does not add useful information. The student should have given a brief explanation of the hydrophobic interaction of the lipid tails and diffusion of lipids in a bilayer.

The section on the mosaic components is correct, but poorly explained. A brief explanation of what types and shapes of proteins are inserted into the membrane would make the answer better.

Essay questions: *Answer 2 of the 3 following questions on the attached blank sheets of paper. Be sure to indicate which ones you are answering! (10 points each)*

1. Describe the major steps of photosynthesis, including the major inputs and outputs.

Answer: Photosynthesis can be broken down into two major parts: a light reaction and a dark reaction. The first part, the light reaction, requires the input of water and sunlight. As sunlight enters, it is first received by photosystem II. The photosystems contain pigments that absorb different wavelengths of light. Photosystem II becomes excited and passes the electrons (which are present as H_2O is oxidized to O_2) down an electron transport chain to photosystem I. The ultimate acceptor of the electrons is the O_2. Through this process, a gradient of H^+ is created between the thylakoid compartment and the stroma. Chemiosmosis drives the reaction as H^+ are driven across the membrane to achieve equilibrium. The ATP molecule that is produced, along with NADPH (which is reduced from $NADP^+$) are the major outputs of the light reaction of photosynthesis.

The major inputs for the dark reaction (Calvin cycle) are the ATP and NADPH produced from the light reaction, along with CO_2. The major output from the process will be G3P, which is used to make glucose. During the Calvin cycle, CO_2 is fixed to form G3P. The ATP is broken down to be used for energy, and the NADPH is oxidized again to NADP+. During the Calvin cycle, various carbon compounds are introduced and formed to create the final output, G3P, for glucose. In addition, it is necessary to recycle the RuBP, so that it can be available to again combine with the CO_2 in a continual cycle.

The answer is well-organized, following the form of the question. It is complete, but not unnecessarily so (an explanation of each intermediate step in the process was not asked for and should not be given). The accuracy of the answer is good, with few flaws in the explanation. The second paragraph is a bit repetitious, and it is clear that the student thought she should be giving more detail, so she repeated herself too much. She clearly stated inputs and outputs, as the question indicated. The writing style was fairly clear and easy to follow.

Many students remember these pathways in a visual form. They may find that it is useful to draw a diagram in the corner or on a sheet of scrap paper to assist in formulating a clear, organized response. For this, as with most questions relating to biochemistry, students should learn these concepts in a visual, dynamic sense. Many of the CD-ROM animated cartoons now available [with textbooks] can help most students envision these processes.

2. Describe the phases of meiosis.

Answer: The first phase of meiosis is interphase I, during which the chromosomes are replicated. (This DNA synthesis occurs during the S phase of the cell growth cycle.) During the second phase, prophase I, the MTOCs with their centriole pairs begin forming the spindle. The nuclear envelope begins to disintegrate, and the nucleoli disappear. During this phase, homologous pairs of chromosomes, along with the sister chromatids, combine to form a tetrad. The passing of genetic information through crossing over then occurs. In metaphase I, the tetrads are aligned in the center of the cell (metaphase plate)—forced there as the spindle fibers connect to the kinetechores. The homologous pairs of chromosomes are then pulled to opposite poles of the cell during anaphase I. Finally, during telophase I and cytokinesis, the cell divides into two.

The second part of meiosis takes place within both resultant cells. Unlike in interphase I, the chromosomes do not replicate in interphase II. The two cells merely undergo their normal growth functions. In prophase II, the MTOCs and centriole pairs again begin forming the spindle in each cell. The nuclear envelope disintegrates, and the nucleoli disappear. The sister chromatids are aligned along the metaphase plate during metaphase II. Spindle fibers again attach to the kinetochores. The sister chromatids are pulled apart during anaphase II to opposite ends. Finally, telophase II and cytokinesis split the two cells into four haploid daughter cells.

This is a straightforward, complete, and well-organized answer. I really could not expect anything better from a freshman!

3. Describe five places in the process of protein construction by a cell at which regulation of gene expression can occur.

The requirement was to respond to two out of three questions. The student chose not to answer this question.

FINAL EXAM FOR GENERAL BIOLOGY II

Write down your questions or points of confusion and see if they are answered during lecture. If not, *ask questions* at the end of class or during office hours. Then reread the chapter for more detailed understanding while taking notes on what you read. Try to apply what you heard and read to things you hear in the news or to the world around you.

***Do not procrastinate* by trying to read chapters for the first time in the day or two before the exam. *Come to class on time,* so you will hear the introductory/context statements for the lecture.**

Short Answer *(5 points each)*

1. Describe the evolutionary relationship between the five major kingdoms of life on Earth.

Answer: Life is thought to have originated on Earth in the form of bacteria. These bacteria, the moneran kingdom, were (and still are) prokaryotic. Over time, membrane infoldings and symbiotic relationships between large and smaller monerans are thought to have created eukaryotic organisms—Kingdom Protista—with a nucleus and other membrane-bound organelles. These eukaryotic protists then developed into the other three kingdoms, fungi, plants, and animals, based on distinguishing characteristics of each group of protists.

The writing is clear with good organization. A short clause on what the distinguishing characteristics mentioned in the last sentence would have helped, however.

2. Give a brief description of the life cycle of a fungus.

Answer: The life cycle of a fungus consists of a dikaryotic, a diploid, and a haploid stage. Haploid nuclei on the underside of the cap of a mushroom (fruiting body) fuse together to form a zygote, the diploid stage. The zygote undergoes mitosis and develops into spores, the haploid stage. The haploid spores fall to the ground, and each grows into a new mycelium. These mycelia fuse together with another mycelium that is complementary to its own nuclei. These nuclei do not fuse, however, so that the new fungus will then contain two haploid nuclei per cell, the dikaryotic stage.

Again, this is one of the well-written and well-organized answers. It is complete, but not excessively long.

3. What features distinguish a plant from an animal?

Answer: A plant has cuticles to protect from loss of water and stomata to allow diffusion of oxygen and carbon dioxide across the leaf. Animals do not. Animals have membrane-bound cells. Plants have cell walls. Plants undergo cellular respiration through photosynthesis. Animals do it chemically. So plants have chloroplasts, animals do not.

This is an example of a poor answer. The information is incomplete, and it is incorrect in several cases or not clearly presented in other cases. The answer implies that plant cells do not have cell membranes, which is incorrect. (They have membranes within walls.) The idea of respiration through photosynthesis is also very strange!

A good answer would contrast the primarily autotrophic plants with heterotrophic animals, presence of cell walls in plants but not in animals, and the greater motility of most animals.

4. Contrast fluid feeders, suspension feeders, and substrate feeders.

Answer: A fluid feeder is an organism that feeds on fluids from another organism. An example would be a mosquito or tick. A suspension feeder is an organism that filters food from around it. Substrate feeders are organisms that either live in or on their food source.

This is a concise and clear answer. When the question asks you to CONTRAST, however, a better answer would not just define each type, but explain the differences between them. A concluding sentence or two could accomplish this.

5. What is the relationship between blood pH and respiration in a mammal?

Answer: When CO_2 levels in the blood get too high, the blood pH increases. There is a direct relationship between pH and respiration.

This is a very poor answer. The first sentence should say *decreases*, not *increases*. The second sentence is merely repeating the question.

A good answer would outline the mechanisms relating blood pH and respiration. It would include a brief discussion of how high carbon dioxide levels lead to reduced pH, an explanation of the role of carbon dioxide sensors in the circulatory system and brainstem, and how those systems lead to an increased breathing rate.

6. Describe the role of the skin in disease fighting.
Answer: Skin is a barrier of stratified squamous epithelial cells. These cells not only provide a layer of dead skin cells to prevent penetration, but they also secrete things. Skin has a low pH to deter foreign organisms from growing and surviving. The skin also secretes cytolytic enzymes that cause bacteria cells to lyse (cuts through their membranes). Sweating also helps skin flush out toxins and helps maintain the unwelcoming pH of skin. Skin is also waterproof due to keratin in cells.

This answer is fairly complete, including the most important points. It is somewhat lacking in organization, with poorly phrased sentences. At one point, the slightly confused writing implies that dead cells "secrete things"!

7. Choose one anterior pituitary hormone and describe its actions on the mammalian body and its regulation.
Answer: The anterior pituitary is very important in the ovarian cycle by secreting the hormones FSH and LH. FSH stimulates the follicles to begin growing. As the follicle grows, it produces estrogen. Low estrogen levels generate negative feedback on the anterior pituitary. As the follicle reaches its largest size, the anterior pituitary releases large amounts of FSH and LH. The FSH has no known effect at this point, but the LH causes ovulation and other associated processes to occur and causes the secretion of estrogen and progesterone from the follicle. These two hormones have the combined effect of exerting negative feedback on the anterior pituitary.

This is a fairly good answer. The negative feedback explanation was a bit confusing and repetitive. Defining the term *negative feedback* would have helped, so would explaining the end result.

The student actually chose one of the more complex hormones to work with. I would suggest that a student presented with an option like this choose a very clear, simple hormone. There is less chance of getting tripped up!

8. Give a brief description of the organization of a skeletal muscle fiber.
Answer: Each skeletal muscle fiber contains many myofibrils. These myofibrils are made up of distinct sarcomeres, which stretch from Z line to Z line. These bands of the sarcomeres are lightly colored at the ends and darker in the middle, where actin and myosin are both found. Muscles contract by a process in which the filaments of the muscle fibers slide over each other and the sarcomeres shorten.

This is a good answer. The last sentence is unnecessary. Many students have difficulty with this question because they confuse a muscle, a muscle fiber, and a myofibril. It is important to know your terms. Rather than discussing the light/dark gradations of a sarcomere, a better answer would have used the terms *thin filaments* and *thick filaments* and associated them with actin and myosin.

Multiple Choice *(2 points each)*

9. Archaebacteria are believed to be the most primitive cells because
a. they are found in extreme environments that may be similar to those of the primitive Earth.
b. they are photosynthetic like ancient plants.
c. unlike eubacteria, they are anaerobes.
d. they have been found in the fossil record.

Answer: a

10. Which of the following are adaptations that help plants live on land?
a. vascular tissue
b. gametangia
c. seeds
d. all of the above

Answer: d

11. You find a vaguely wormlike animal in a mudflat. It is bilaterally symmetrical, is segmented, has a coelom, and has a complete digestive tract with the anus opening at the tip of the body. Which of the following choices correctly pairs one of these characteristics with a phylum it rules out?
a. Complete digestive tract rules out Annelida.
b. Coelom rules out Platyhelminthes.
c. Segmentation rules out Chordata.
d. Bilateral symmetry rules out Mollusca.

Answer: b

12. Which of the following is NOT a true statement about organs?

a. An organ represents a higher level of structure than the tissues composing it.
b. An organ consists of several tissues.
c. Organs are found in virtually all animals except sponges.
d. Organs are limited to one function.

Answer: d

13. Digestion takes place in specialized compartments for all of the following reasons EXCEPT

a. the environment of digestion must protect the food.
b. the environment of digestion must favor the action of digestive enzymes.
c. the specialization of the sites of digestion promotes efficiency.
d. the animal's body must be protected from its own digestive enzymes.

Answer: a

14. Oxygen is mostly transported through the mammalian body

a. dissolved in the blood.
b. dissolved in red blood cells.
c. bound to hemoglobin.
d. bound to dissolved iron.

Answer: c

15. Which of the following factors does NOT contribute to the flow of blood in veins?

a. skeletal muscle contraction
b. blood pressure generated by the heart
c. gravity
d. one-way valves

Answer: b

16. A researcher detects interferon in a laboratory rat and concludes that

a. one or more neighboring cells were infected with a virus.
b. cancerous cells are present in the rat.
c. the rat's diet is deficient in calcium.
d. the complement system is activated by exposure to an antigen.

Answer: a

17. What type of immune response is always disadvantageous to a person?

a. cell-mediated

b. inflammatory
c. humoral-mediated
d. autoimmune

Answer: d

18. Which of the following animals is warming its body by behaving as an endotherm?
a. a beetle that absorbs solar radiation
b. a moth that shivers its wings before flight
c. a lizard that lies on a warm rock
d. a turtle that move to a warm, shallow part of a pond

Answer: b

19. Which of the following classes of nutrients produce the largest amounts of toxic compounds when broken down during metabolism?
a. carbohydrates
b. fats
c. cholesterol
d. proteins

Answer: d

20. What component of the renal system is permeable to urea?
a. proximal tubule
b. collecting duct
c. distal tubule
d. loop of Henle

Answer: b

21. A cell that is affected by a particular steroid hormone would be expected to have
a. DNA sites that interact directly with the hormone.
b. an intracellular receptor protein that binds the hormone.
c. a cell-surface receptor protein that binds the hormone.
d. enzymes that are activated or inactivated by the hormone's second messenger.

Answer: b

22. Why are human testes located in an external sac rather than in the abdominal cavity?
a. to shorten the distance which semen must travel during ejaculation
b. to shorten the distance which sperm must swim during insemination
c. so the testes can be kept at a constant temperature

d. so the testes can be kept cooler than the body's interior

Answer: d

23. What does gastrulation accomplish?

a. It changes a solid embryo into a hollow morula.

b. It changes the solid blastula into a hollow embryo that has three tissue layers.

c. It changes the hollow blastula into a hollow embryo that has three tissue layers.

d. It creates the neural tube by invagination of the ectoderm.

Answer: c

24. Sodium is removed from a nerve cell

a. by diffusion.

b. by osmosis.

c. during an action potential.

d. by active transport.

Answer: d

25. Which of the following choices INCORRECTLY pairs a class of sensory receptor with one of the kinds of stimulus it detects?

a. chemoreceptors—molecular structure

b. photoreceptors—infrared radiation

c. nociceptors—pressure

d. mechanoreceptors—sound

Answer: c

26. Muscle contraction causes

a. only the thin filaments to shorten.

b. only the thick filaments to shorten.

c. both thick and thin filaments to shorten.

d. only the sarcomere to shorten.

Answer: d

27. Which part of a bone contains stored fat?

a. yellow bone marrow

b. fibrous connective tissue

c. cartilage

d. spongy bone

Answer: a

28. Countercurrent exchange in the gills of a fish

a. speeds up the flow of water through the gills.

b. interferes with the efficient absorption of oxygen.

c. maintains a gradient that enhances diffusion.

d. enables the fish to obtain oxygen while swimming backward.

Answer: c

Essay: *Choose two of the three questions to answer. (10 points each)*

29. Describe the conditions on Earth during the formation of the first life-forms. Describe the major molecular events necessary to the formation of the first cells.

The student did not chose this question.

30. What is homeostasis? What are the advantages and disadvantages of being a conformer vs. a regulator?

Answer: Homeostasis is the ability of an animal to control its internal environment. Animals fall into two types of categories. They are either conformers or regulators. A regulator keeps tight control on such things as body temperature, salt content, and body pH. The advantages of being a regulator are that you do not have to worry about becoming too hot or cold for mobility or that your enzymes might not work because you get hot enough to denature them. A conformer on the other hand does not regulate much of its body systems. A disadvantage of being a conformer is that they cannot survive well in extreme conditions. However, an advantage is that you do not have to find as much food. Compared to a conformer, regulators eat more per body size to keep themselves fueled.

This answer contains the major points, although the definition of homeostasis could use a bit more explanation and can apply to no-animals in some cases as well. It is not surprising, however, that the student presented it only for animals, because that is how the textbook presented the concept. I always dislike answers written with "you," as if the reader were necessarily the creature being used as the example. This student needs some work on writing, but has the correct concepts.

31. Describe the pathway of blood movement through the mammalian cardiovascular system, including pressure and velocity changes.

Answer: Blood enters the right atrium of the heart via the inferior and superior vena cava. As blood traveled through the veins to the heart, blood pressure dropped to zero, but velocity increased from the slow speed the blood traveled as it passed through the capillaries in the systemic circuit. Blood enters the left atrium via the pulmonary vein. Pressure there is also low, but velocity increased after the pulmonary circuit. The velocity of blood is always lowest in the capillaries, and the blood pressure is always lowest (almost zero) in the veins. The sinoatrial node in the heart generates an electrical signal that travels to the atrioventricular node. Blood flows from the left atrium into the left ventricle and from the right atrium into the right ventricle as the AV valves open. Then, blood is forced out of the semilunar valve of each respective ventricle. As blood leaves the left ventricle (is pumped out), pressure and velocity are both very high. The blood travels to the lungs via the pulmonary artery, slows down in the systemic capillaries as oxygen is picked up, and then increases in velocity in the pulmonary vein (low pressure) before entering the left atrium. Blood is pumped much harder from the right ventricle so it can reach all parts of the body. Velocity and pressure are both at their highest as blood leaves the heart via the aorta. The blood velocity slows in the systemic capillaries as oxygen is given up and carbon dioxide is picked up. Velocity then increases as blood flows back to the right atrium via the vena cavae, but pressure diminishes to almost zero. The cycle then repeats.

This is a correct, but poorly organized and repetitive answer. The student would have done better first to describe the pathway, beginning with the entrance of blood into the right atrium and going through the entire circuit to return to the right atrium, then address the pressure and velocity changes. Or the student could have put the sentence beginning "The velocity of blood is always lowest . . ." as a concluding sentence and just put in parentheses, after each step in the process, an indication of whether pressure and velocity are high or low at that point.

CHAPTER 28

VASSAR COLLEGE

BIOL 152: THE ROLE OF THE CELL

Leathem Mehaffey, III, Associate Professor

TWO "MIDTERM" EXAMS AND ONE FINAL ARE GIVEN IN THIS COURSE. THE MIDTERMS each count for 15 percent of the final grade in the course. The final counts for 20 percent. One midterm and the final are included here.

Primary course objectives are to teach students some basic principles of biology using cells as an example system. We as a department some years ago discussed the basic principles we wished students to learn and how best to convey them. We considered introductory courses using the cell, ecology, or physiology as the vehicle, and we decided on the cell.

I expect students to come out of this course with a feeling for the kinds of problems faced by all living systems (energy acquisition, reproduction, adaptation—both immediate and long-term, i.e., evolution). Since we use cells as the vehicle for transmitting this information, I expect students to know some basic cell physiology concepts such as the advantages of regional specialization as reflected in compartmentalization; enzyme kinetics and how it pertains to cells' regulation of product and reactant concentrations and feedback loops; the structure, reproduction, and maintenance of DNA; the basic dogma of DNA-RNA-protein and how this is controlled in an adaptive way; and cell cycling and differentiation.

The questions asked attempt to deal with the implications of the facts rather than just the facts themselves. However, as in any new subject, one must learn the vocabulary and some basic facts. But, in this course, students are asked to go beyond the facts to the implications of those facts for living systems, and to apply what they have learned about cells to biological systems in general. I tell them to focus on those facts that give the broader implications, such as adaptive regulation of metabolic pathways.

HOUR TEST 2

1. [10 points] Identify the specific cellular, mitochondrial, or chloroplast location for each of the following:

a. The enzymes of glycolysis.
Answer: Cytoplasm

b. The electron transport chain in oxidative phosphorylation.
Answer: Inner membrane of mitochondrion

c. The enzymes of the Calvin cycle.
Answer: Stroma of the chloroplast

d. The electron transport chain in photosynthesis.
Answer: Thylakoid membrane of the chloroplast

e. Docking proteins
Answer: Membrane of the specific organelle, e.g., surface of the endoplasmic reticulum, mitochondrion outer membrane, etc.

I advise students to calculate the time based on the fact that they have 50 minutes for a 'midterm' exam and 120 minutes for a final; in each case, the exams are worth 100 points. So a 10-point question should consume no more than 10% of the time, or 5 minutes for an in-class exam or 12 minutes on a final.

2. [15 points] The Pasteur effect states that a cell in the absence of oxygen runs glycolysis much faster than it does if oxygen is present. Explain.
Answer: Glycolysis is the process by which cells oxidize glucose or glycogen in order to obtain energy. In the process of oxidation electrons are removed from the compound. In order for this to happen something must accept the electrons and be reduced. If oxygen is present, then the oxygen becomes the acceptor and is reduced to water. Under these circumstances the glucose is fully reduced to CO_2 and H_2O, producing about 30 moles of ATP per mole of glucose and utilizing the electron

transport system of the mitochondria. When oxygen is not present ATP can still be obtained, but another electron acceptor must be found, namely NAD. However, this shortened path yields only 2 moles of ATP, not 30. Since glycolysis is regulated by negative feedback of ATP on the enzyme phosphofructokinase (PFK), the lowered production of ATP causes speedup in the reaction catalyzed by the PFK and hence a speedup in glycolysis.

This answer goes a bit more into background than needed, but that makes for a more understandable conclusion.

The question is designed to make the student use his or her knowledge of regulation in cells. Approach the question by thinking about the phenomenon of the Pasteur effect. You know that glycolysis can run aerobically or anaerobically. You also know that the anaerobic pathway produces much less ATP. Connect this with the knowledge that PFK, the major regulatory enzyme of glycolysis, is negatively regulated by ATP.

3. [15 points] SHORT ANSWERS: Answer ANY FIVE.

a. What are at least two functions of the Krebs cycle?
Answer: production of energy (NADH, ATP) and interconversion of metabolites

b. What is meant by a "redox pair"?
Answer: a pair of reactions in which one acts as an electron donor and the other as an electron acceptor

c. Differentiate between substrate level and oxidative phosphorylation.
Answer: Substrate level: direct transfer of high-energy phosphates from a metabolite to ADP.

Oxidative phosphorylation: use of the electron transport chain to produce a proton gradient, which in turn is used to produce ATP.

d. List three organelles bounded by double membranes.
Answer: nucleus; mitochondrion; chloroplast

e. What are two functions of microtubules? Of intermediate filaments?
Answer: Microtubules: motility (cilia, flagella), intracellular transport

Intermediate filaments: structure (nuclear lamina), stress handling (cell-cell connections such as desmosomes)

f. Match the motor molecule with the appropriate filament:
Answers are in italics:

Dynein: *microtubules*
Kinesin: *microtubules*
Myosin: *actin*

4. [10 points] We have said that glycolysis can be run both forward and backward, either to degrade or to produce glucose. Is this true for every reaction in the pathway? On what do you base your answer?

Answer: This would not be true for every reaction in the pathway. The reactions have very different Gibbs free energies. ΔG is a measure of the difference between the actual ratio of product to reactant in the cell and the ratio at equilibrium. Reactions with small ΔG can be easily reversed by altering the ratio of product to reactant since the ratios are near to equilibrium anyway. Reactions with large ΔGs are far from equilibrium and so one cannot easily alter the ratio enough to reverse the reaction.

The response clearly states the first answer as no. It then goes on to explain the background of that answer. Students must know enough about Gibbs free energy to know that it is dependent on the ratio of product to reactant as found in the cell, and that large ΔGs mean ratios far from equilibrium. They must also know that to reverse a reaction you have to change the sign of ΔG.

Analyze the question. Students should know from lecture and readings that some reactions in glycolysis are not reversible and have to be bypassed in gluconeogenesis. Then make the connection between the ΔGs of these reactions and their reversibility. That plus a knowledge of what ΔG has to do with product—reactant ratios and why cells maintain those ratios—should lead to an answer.

5. [10 points] How can forensic (police) scientists use rigor mortis to determine the time of death of a murder victim?

Answer: Rigor mortis means the stiffening of a body due to rigidity of the muscles, including opposing muscles. This rigidity is related to actin-myosin cycling within the muscles. The myosin heads cannot come off the actin except in the presence of ATP. Thus, following death, when the ATP supply runs out, rigor mortis sets in and lasts until the proteins begin to degenerate. Forensic scientists can estimate the initially available ATP supply and external factors such as temperature and thus use the presence of rigor mortis to estimate the time of death.

This answer shows a clear flow of logic to a conclusion. To answer successfully, the student must know what rigor mortis is and must know enough of the sliding filament theory to assess its relevance to muscle stiffness. The student also has to know the place in the actin-myosin cycle where ATP comes in.

A good approach to the question is to begin by asking what rigor mortis is—muscle stiffness. What did you learn about muscles? Sliding filament. What in that theory would explain muscle stiffness? Sustained contraction? No, you'd soon run out of ATP. Oh, that's it. The ATP is needed to cause the myosin to lose its affinity for the actin, so in the absence of ATP the muscles would be stiff. Dead people don't make ATP, so rigor would set in after the already-existing supply of ATP was used up. And that gives you a kind of clock.

6. [10 points] What factors influence the polymerization of microtubules?

Answer: concentration of tubulin; hydrolytic activity of the tubulin for its bound GTP; the activity of microtubule associate proteins

7. [5 points] List three roles of actin in cells.

Answer:

1. muscle contraction
2. filopodia extension
3. cytokinesis

8. [10 points] Explain the idea of "end-product inhibition" and how it acts to stabilize concentrations in the cell.

Answer: End-product inhibition means the inhibition of a synthetic reaction (or series of reactions) by one (or more) of the end products of that reaction. It stabilizes concentrations in cells in that when the concentration of product rises the reaction(s) producing it slow down, and when the product concentration drops the reaction(s) producing it speed(s) up. This will tend to stabilize the concentration of the product even when the utilization of the product by the cell varies over time, or the concentration of the precursors varies.

The main issue here is control of concentration by the cell. A second issue is the need for such control.

9. [15 points] DATA ANALYSIS. Using the data provided, answer the question. You have available to you the following chemical and pharmacological information about the intermediaries in the electron transport chain. Use the information to determine the sequence of the intermediates and the points in the chain where there might be sufficient energy change to produce ATP. *Explain your reasoning.*

I. redox potentials:

$NADH + H^+ \rightarrow NAD^+ + 2H^+ + 2e^-$	-0.32 V
cytochrome $c_{ox} + e^- \rightarrow$ cytochrome c_{red}	+0.25 V
cytochrome $a_{ox} + e^- \rightarrow$ cytochrome a_{red}	+0.29 V
cytochrome $a_{3(ox)} + 2e \rightarrow$ cytochrome $a_{3(red)}$	+0.39 V
coenzyme $Q_{ox} + 2H^+ + 2e^- \rightarrow$ coenzyme Q_{red}	+0.10 V
cytochrome $b_{ox} + e^- \rightarrow$ cytochrome b_{red}	+0.06 V
$½O_2 + 2H^+ + 2e^- \rightarrow H_2O$	+0.82 V

II. pharmacology and biochemistry:

If $NADH + H^+$ is used as a precursor (entry way into the chain), 3 ATP are produced. If coenzyme Q_{red} is used as the precursor, 2 ATP are produced. If cytochrome c_{red} is used as the precursor, only 1 ATP is produced.

Answer: The sequence is NAD, cytochrome b, coenzyme Q, cytochrome c, cytochrome a, cytochrome a_3, and O_2. This is based on knowledge that NADH is the entry point from glycolysis, and O_2 is the final acceptor. Since NAD has the most negative redox potential and O_2 the highest, clearly things with more positive potentials can oxidize things with more negative potentials. Thus the other reactions should fall in sequence from negative to positive.

One ATP must be produced between NAD and coenzyme Q, since entry with coenzyme Q produces two ATP but entry with NAD produces three. Another ATP must be produced between coenzyme Q and cytochrome c, by the same logic. The third ATP must be produced between cytochrome c and oxygen. To be more specific, it appears that the ADPS would be produced at the redox pairs with the biggest difference in redox potential, since ΔG is related to redox potential. Thus I would guess the first ATP is produced at the NAD-cytochrome b reaction (ΔE = -0.32 – 0.06 = –0.38 V, whereas for cytochrome b-coenzyme Q, ΔE = 0.06 – 0.10 = –0.04 V). The next is at the cytochrome q-cytochrome c reaction, which has a ΔE of –0.15 V. The last is probably at the cytochrome a_3 – O_2 reaction, since it has a ΔE of –0.43 V, whereas the other reactions between cytochrome c and oxygen have ΔEs of –0.04 and –0.10 V.

The answer shows a logical progression building from the facts to the conclusion. Students need not have memorized the electron transport chain to answer this one, but they do need to know that redox reactions are paired. They also need to understand the relationship between redox potential and free energy, and that the electron transport chain is a series of redox reactions.

To answer successfully, look first at the redox potentials. These are each for half-reactions, not redox pairs. You should know that all redox reactions must occur in pairs, with an electron donor and a receptor. Knowing the start point (NADH) and the end point (H_2O), you should be able to place the half-reactions in a sequence.

Once that's done you can tackle the second part of the question. A sketch might help show where the ATPs are produced, at least between which reactions (e.g., NAD and coenzyme Q). If you know that ΔE of a redox pair is related to ΔG, you can make a pretty good guess as to which reactions

have enough energy to produce an ATP, especially given that you know that an ATP is produced between the coenzyme Q and cytochrome c reactions where the ΔE is −0.10 V.

FINAL

Of course, the number one thing is to stay up with the course in terms of readings and lecture attendance. Read text material before coming to class and ask questions in class. As to studying for the exams, work in groups and ask each other questions; don't just memorize facts.

IMPORTANT NOTE: Questions have different weights. Budget your time accordingly.

1. [6 points] Label the following as guanine, thymine, ribose, progesterone, deoxyribose, or amino acid.

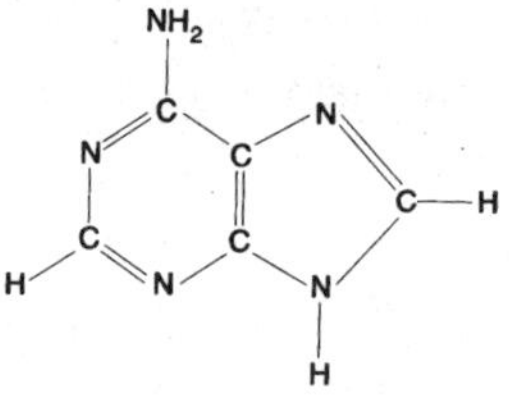

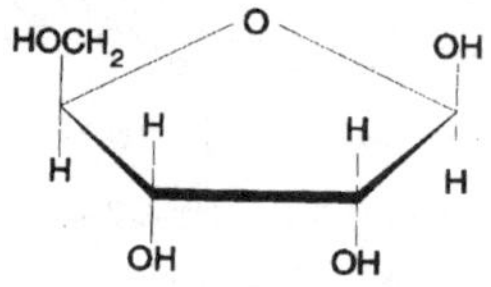

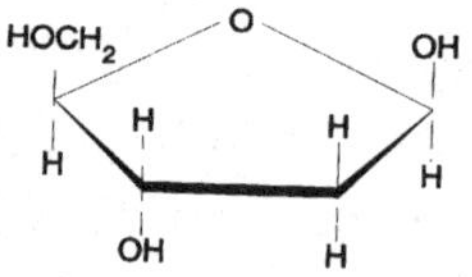

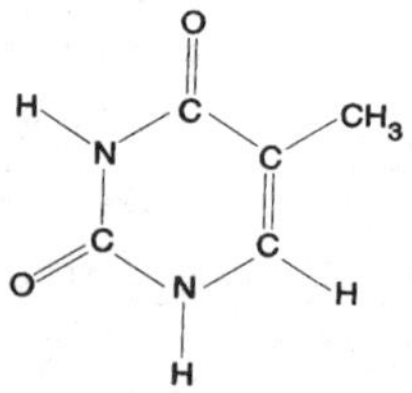

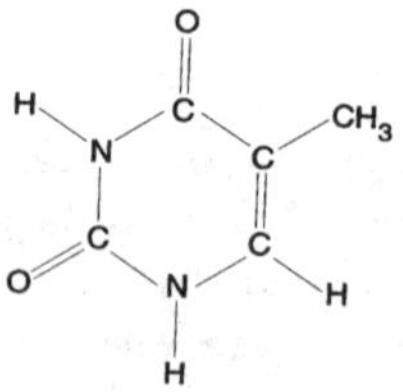

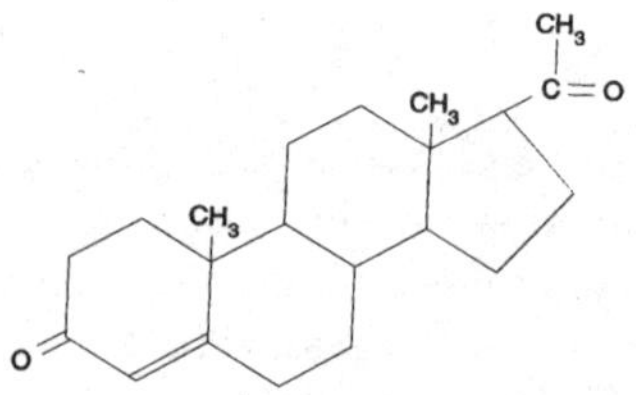

Answers:
Left to right, row 1: guanine, ribose
Left to right, row 2: deoxyribose, thymine
Left to right, row 3: amino acid, progesterone

2. [5 points] Why does DNA polymerase work in one direction only? What is that direction?

Answer: DNA polymerase works only in the 5′→3′ direction because in order to form the bond between the next nucleotide and the end of the DNA molecule it must use the energy from the high-energy phosphate bonds on the incoming 5′ triphosphate nucleotide, which it does by hydrolyzing that triphosphate to a monophosphate. Thus the bond is formed between the 5′ end of the new nucleotide and the 3′ end of the existing strand, the remaining phosphate acting as the bridge. This process would not work from the 3′→5′ direction unless 3′ triphosphate nucleotides were used, and these do not normally occur.

This answer is concise yet complete. It also considers the only possible alternative and accounts for that. The main issue implicit here is the understanding of the economical use of energy by the cell and the fact that enzymes work on a molecular level, here meaning that an enzyme designed to work on 5′ triphosphonucleotides will not recognize 3′ ones.

Approach the question by first analyzing it. First, the question involves enzymes, so the answer may well lie in the properties of enzymes. Second, it is a synthetic reaction, so it requires energy. Where does the energy come from?

3. [6 points] Compare the organization of the genome of eukaryotes and prokaryotes as to:

a. Overall structure

Answer: Prokaryotic DNA is circular; that of eukaryotes is linear. The procaryotic DNA is supercoiled and probably associated with some proteins that further coil it. Eukaryotic DNA is wrapped around nucleosomes, and then the DNA-nucleosome strand coils into solenoids, which may again coil into "supersolenoids."

This answer is short and factual. The question gets to the packing of the large amount of DNA needed to contain the information for a cell into the small volume of a cell. It emphasizes the difference between eukaryotes and prokaryotes, which difference was emphasized in lecture in an evolutionary context. The best approach is to keep the answer simple and straightforward.

b. Organization of genes on the DNA

Answer: Prokaryotic genes are organized in cassettes or operons with a series of structural genes along with their associated regulatory sites (promoter, etc.). This is called "polycistronic." Eukaryotic genes are monocistronic, with each single gene associated with its promoter. Regulatory sites can be many base pairs away from the structural gene. Eukaryotic genes often contain introns, or portions of the DNA that are not translated.

4. [5 Points] Telomeres have been linked to aging; in fact, using genetic engineering to

maintain the activity of telomerase has been called "the fountain of youth" or even a clue to immortality.

a. What do telomeres do?

Answer: Telomeres provide a section of DNA at the end of the eukaryotic linear chromosome, which allows the DNA polymerase enzyme to completely replicate the lagging strand. This is necessary since the lagging strand can only be synthesized in the 5′→3′ direction and so there must be a place beyond the last gene on the 5' end of the lagging-strand DNA for the polymerase to work from.

Students must know how DNA polymerase works and associate this with the linearity of DNA in eukaryotes.

b. Why would restoring telomerase activity potentially lead to immortality of the whole organism?

Answer: Telomerase makes the telomeres and restores them as they are used up in DNA replication. Without telomerase the telomeres get shorter at each DNA replication until they are gone and the true coding DNA on the 5' end of the lagging strand is exposed. Thereafter a little of this will be lost at each replication. Restoring telomerase activity would prevent this and allow the organism's cells to replicate forever.

Students are asked here to use the knowledge they have acquired in a novel way, and to combine their knowledge of telomeres with their knowledge of why organisms senesce.

5. [8 points] Below is a diagram of a ribosome in three stages of protein synthesis. On it, draw in a messenger RNA, a transfer RNA carrying an amino acid, the growing polypeptide, and an empty transfer RNA leaving the ribosome.

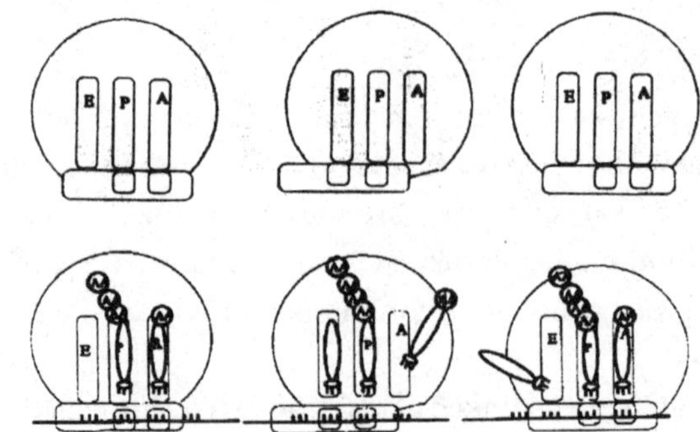

Answer:

6. [5 points] Brenner and Crick in 1961 found that deletion of one base pair in a DNA molecule led to production of abnormal, nonfunctional proteins. The same was true for two-base-pair deletions. Would the same be true for three-base-pair deletions? Five-base-pair deletions? Six? Explain your answer.

Answer: Three-base-pair deletions should often lead to a normal protein, since three base pairs is the codon for one amino acid, the loss of which will not usually damage the protein. One- or two-base-pair deletions cause a "frame shift" in which all subsequent codons read by RNA polymerase are nonsense. For example, UAG-GAC-AGG becomes UGG-ACA-GG by loss of the A in the first codon. The same kind of thing would happen for any deletions that are not multiples of 3; so a four- or five-base-pair deletion would be bad, whereas a six-base-pair deletion might be okay.

The answer explains the phenomenon stated in the question, then uses that knowledge to make the predictions asked for. Students must know that DNA codes for amino acids in proteins and that each codon is a three-base-pair sequence. They should also know that altering one amino acid need not result in a nonfunctional protein.

Think about the experimental result being described and use your knowledge of DNA to reason out the answer.

7. [8 points] In the following table, give an example of control of gene expression at each indicated level and some advantages and disadvantages of exercising control at that level as opposed to the other levels.

Answer: The answers are shown in italics.

Control Level	Example	Advant./disadvant.
Genomic	*heterochromatin/ euchromatin*	*A: long-term, many genes.* *D: slow*
Transcription	*repressors/enhancers*	*A: faster, widespread effects* *D: timespan of hours*
Post-transcription	*mRNA processing*	*A: same transcript, many mRNAs.* *D: wasteful of RNA, timespan of hours*
Translation	*initiation, elongation, termination factors*	*A: fairly fast, e.g. phosphorylation. Targeted to regions of cell.* *D: action is local*
Post-translation	*phosphorylation*	*A: very fast (msec). Very local.* *D: very local*

8. [5 points] What is meant by "combinatorial control" in eukaryotes? How is it different from prokaryotic gene regulation?

Answer: In eukaryotes a single promoter is not usually sufficient to cause a gene to be transcribed at its maximal rate. A number of factors are required. This allows for graded response all the way from no transcription to transcription at the maximal rate, depending on which regulatory factors are activated. Also, the same gene can be controlled by several external factors working through different regulatory proteins on the one hand, or different genes can be controlled in different cells by the same external factors working in concert with other regulatory proteins specific to each cell.

This question gets to the heart of the concept of regulation and response of the cell to different environmental cues, as well as differential responses of cells.

9. [5 points] Unequal crossing over can lead to gene duplication. When in the life of a cell might this occur? What is the significance of this for evolution?

Answer: Unequal crossing over can occur whenever homologous chromosomes are bound together in synapsis. In most eukaryotes this will occur only during meiosis. Its significance is that it can lead to duplication of genes, allowing the duplicated gene to mutate repeatedly without deleterious effects due to deletion of a function, which in turn can lead to speeding up evolution.

This question is central to the understanding of evolution on a molecular scale.

10. [5 points] Describe how cyclins and cdk proteins control the cell cycle. What role does protein p53 play in this cycle? What role does ubiquitin play?

Answer: Both cyclin and cdk are produced constitutively by the cell throughout its cycle. cdk remains at a constant level, but the concentration of cyclin, on the other hand, rises steadily through interphase and then drops precipitously after mitosis. When the concentration of cyclin reaches a threshold value, the cdk and cyclin combine to make an active kinase called mitosis promoting factor or MPF. This kinase phosphorylates a number of proteins involved in mitosis and brings about the onset of mitosis. The cyclin is bound to ubiquitin, and the activation of the MPF appears to activate the ubiquitin tag on the cyclin, causing it to be destroyed by the ubiquitin-dependent proteolytic system.

The major issue here is regulation.

11. [5 points] Describe how a G-protein conveys a signal from a cell-surface receptor to some intracellular effector. How is the system restored to its original state?

Answer: A ligand binds to a receptor (typically a seven-pass transmembrane protein), which then on its intracellular side becomes activated. Upon the next collision between a G-protein and this site the G-protein trimer ($\alpha\beta\gamma$) exchanges its bound GDP for a GTP on the α subunit and dissociates into the α-GTP subunit and the $\beta\gamma$ subunit. The α-GTP subunit travels along the inside surface of the membrane. When it collides with an effector such as phospholipase C, it can activate that effector. Restoration of the system is effected by the hydrolytic activity of the α subunit, which hydrolyzes the GTP to a GDP making the subunit inactive and restoring its affinity for the $\beta\gamma$ subunit.

The major issue here is the transmission of information from the outside environment of the cell to its interior.

12. [5 points] Why are the ligands for receptor tyrosine kinases always dimers?

Answer: Because the receptor tyrosine kinases are single-pass transmembrane proteins, two of which must be brought together for mutual phosphorylation before they can become active. Dimeric ligands bind two such receptors and hold them in proximity to each other.

13. [5 points] What is the logic behind the dideoxyribonucleoside method of DNA sequencing?

Answer: One uses four reaction vessels. In each are all the raw materials for DNA synthesis, including DNA polymerase, a template DNA, and an assortment of triphosphodeoxynucleotides. In one vessel there is a small amount of dideoxyATP; in another, dideoxyGTP; in a third, dideoxyCTP; in the fourth, dideoxyTTP. In the first vessel the various polymerization reactions will proceed until each one tries to use one of the dideoxyATPs, at which point the reaction will stop since the next nucleotide can't bind to the dideoxyA. After a while you have a number of fragments, each of which ends with an A on its 3′ end. In the second vessel the same is true except that all fragments end with a G, and so on for the third and fourth vessels. Running the DNA from each vessel in one lane of a gel allows one to read up from the bottom of the gel (shortest fragment) to the top (longest fragment), reading across the lanes to get the sequence that was in the original (template) DNA.

A successful response requires understanding of the use of DNA technology.

14. [5 points] What is the advantage of cellular compartmentalization, and how do the properties of membranes operate to confer this advantage on eukaryote cells?

Answer: Cellular compartmentalization allows for intracellular specialization, which allows specific portions of the cell to carry out specific tasks most efficiently in that they contain all the necessary enzymes and substrates for a set of tasks. Cellular compartmentalization also allows for the maintenance of different concentrations of reactants in different parts of the cell, which permits the Gibbs free energy of a reaction to be tailored to the needs of the special locations in a cell. Thus the hydrolysis of ATP might be favored in one location but the synthesis of ATP be favored in another. Eukaryotic cell membranes are phospholipid bilayers and hence do not allow the passage of hydrophilic substrates and proteins from one compartment to another. The membranes also act as matrices for membrane-bound proteins, anchoring them to one surface or even to a particular location on that surface.

The major issues here are intracellular regulation, specialization, and efficiency.

15. [6 points] Complete the following table:

Answer: The answers appear in italics.

"Motor molecule"	Substrate Protein	Preferred Direction
Myosin	*Actin*	*minus to plus*
Kinesin	*microtubules*	*minus to plus*
Dynein	*microtubules*	*plus to minus*

16. [5 points] What do the K_m and V_{max} tell you about an enzyme? Are these universal properties of that enzyme?

Answer: K_m tells you the substrate concentration needed for the enzymatic reaction to achieve half its maximal velocity. This tells you about the affinity of the enzyme for the substrate: smaller K_m implies greater affinity. The V_{max} of the reaction is the maximum velocity at which the reaction will proceed regardless of any increase in substrate concentration. These are not universal properties of the enzyme but only of the specific reaction. Raising the concentration of the enzyme will increase V_{max}. Competitive inhibitors will increase K_m. Noncompetitive inhibitors will decrease V_{max}.

This is a question about how enzymes work. This question follows up on the determination of V_{max} and K_m done in the lab.

17. [6 points] Complete the following table for photosynthesis:
Answer: Answers appear in italics in the table.

Reaction complex	Input(s)	Output(s)
Photosystem I	*electrons from photosystem II; light*	*NADPH*
Photosystem II	*electrons from H_2O; light*	*high-energy electrons; H^+ pumping*
"Dark" reactions	*NADPH, ATP, CO_2*	*NADP, ADP, PGAL*

18. [5 points] How is the Gibbs free energy (ΔG) of a reaction related to the concentrations of the reactants and products in a cell? What does this tell you about the importance of end-product inhibition?

Answer: The Gibbs free energy available from a reaction (or needed by the reaction) is equal to the standard free energy ($RT \ln K_{eq}$, where K is the equilibrium constant or ratio of product to reactant at equilibrium) plus $RT \ln [P]/[R]$, where $[P]$ and $[R]$ are the actual concentrations of reactant and product in the cell at the site of the reaction. End-product inhibition is used to regulate the concentration of the product of a reaction, and hence it regulates the Gibbs free energy from that reaction. It can thus keep a reaction moving in the forward (or backward) direction as the cell needs.

Regulation within the cell is the central issue in this question.

PART FOUR

FOR YOUR REFERENCE

CHAPTER 29

A GLOSSARY OF BIOLOGY

Abiotic Nonliving.

Acoelomate An organism without a body cavity.

Acquired characteristics Adaptations or other changes that occur during an organism's lifetime.

Adaptations Characteristics and features that contribute to an organism's survival and success in an environment.

Adenosine triphosphate (ATP) An energy-rich molecule important in metabolic processes.

Aerobic Requiring oxygen or describing a condition in which oxygen is present.

Algae Water-dwelling eukaryotes with plastids and without embryos.

Allele One of the possible variants of a given gene.

Alternation of generations Life cycle in which gametophyte and sporophyte stages alternate.

Amino acid An organic compound containing an amino group ($—NH_2$); a building block for protein.

Anaerobic Not requiring oxygen or describing a condition in which oxygen is not present.

Angiosperm Flowering seed plant.

Anterior Front end of an animal exhibiting bilateral symmetry.

Antibody A protein released in response to an antigen to fight infection in an organism.

Anticodon A three-nucleotide sequence of tRNA; pairs up with a codon.

Antigen Substance provoking an immune response.

Arboreal Describes organisms living in trees.

Archenteron Precursor, in the gastrula, of the mature gut.

Artificial selection In contrast to natural selection, genetic selection by purposeful, artificial means, as in cultivated, domesticated species.

Autosomal A trait that is *not* sex-linked.

Autotroph Organism requiring no external food source; derives energy from sunlight.

Bilateral symmetry Mirror-image symmetry typical of relatively advanced animal organisms.

Binomial nomenclature Two-word naming of species.

Biome Large grouping of ecologically related communities with ecologically related species.

Biosphere The Earth and all the organisms of the Earth.

Blastula Developmental stage in which the embryo is a hollow ball of cells.

Blood Fluid tissue containing erythrocytes, leukocytes, and platelets.

Calvin cycle Cycle of chemical reaction in which CO_2 is incorporated into a 3-carbon compound.

Carbohydrates Sugars, starches, and related food compounds.

Carnivores Meat-eating animals.

Catalyst A substance that accelerates a chemical reaction without being used up or changed in the process.

Cell The smallest unit of an organism. Contained within a cell membrane, the cell contains DNA and RNA, and is the site of the fundamental life functions.

Cell membrane Outer living lining of a cell; also called the plasma membrane

Centriole Microtubules that form spindle fibers during mitosis.

Centromere The central portion of a chromosome; it holds the chromatids together.

Chitin The protein-polysaccharide substance that is the primary constituent of exoskeletons.

Chloroplast Plant organelle bearing chlorophyll.

Chordata The phylum that includes animals with notochords.

Chromatid A chromosome strand.

Chromatin The DNA-bearing material that is organized into chromosomes within the cell nucleus.

Chromoplast A plant-cell organelle containing pigments other than chlorophyll.

Chromosomes The cellular structures that carry genes.

Cilia Hairlike projections on some cell surfaces, used in locomotion and food-gathering.

Codon A three-nucleotide mRNA genetic message.

Coelom A fluid-filled body cavity lined with mesoderm.

Cofactor Nonprotein component of some enzymes.

Cold-blooded Describes an animal whose body temperature is close to that of the external environment.

Community The species that interact within a particular habitat.

Covalent bond Chemical bond formed by shared electrons.

Crossing over In prophase of meiosis I, the breakage, switching, and rejoining of homologous chromosomes.

Cytokinesis After mitosis, division of cell cytoplasm.

Cytoplasm All of the cell outside the nucleus and within the cell membrane.

Cytoskeleton Framework of a cell, including microtubules and microfilaments.

Diploid Having two chromosomes of each type; the number of chromosomes in somatic cells nuclei.

DNA Deoxyribonucleic acid A nucleic acid containing deoxyribose sugar, DNA carries genetic information and is capable of self-replication as well as synthesis of RNA.

Dominant An allele that produces the same phenotype whether inherited with a homozygous or heterozygous allele.

Ecosystem A community *and* its abiotic surroundings.

Ectoderm The outer layer of cells.

Egg The female gamete. Also called an ovum.

Electrophoresis Laboratory technique that identifies proteins by measuring their speed of migration through an electrical field.

Embryo Early stages of an organism's development.

Endocrine Refers to ductless glands that secrete substances into the bloodstream, which affect a target organ or organs.

Endoderm Inner layer of cells.

Endoplasmic reticulum Structure within cell cytoplasm that transports materials within the cell.

Endosperm Seed tissue containing stored food.

Enzyme Organic catalyst composed mainly of protein.

Epidermis Outer, protective cell layer.

Epithelium Tissue originating in broad, flat surfaces.

Erythrocytes Red blood cells.

Eukaryotic Describe a cell with an organized, membrane-bound nucleus and organelles.

Exocrine Describes glands that secrete into a duct or along a surface rather than into the bloodstream.

Exoskeleton An external skeleton.

Extinction Death of a species.

Falsifiable Capable of being proven false; required of a hypothesis in the scientific method.

Fatty acids Long-chain molecules containing —COOH at one end.

Fermentation Anerobic metabolism; produces carbon dioxide.

Fertilization Union of egg and sperm; produces a zygote.

Fetus Stage in mammalian development beyond embryo, in which organs are formed.

Flagella Whiplike organelle on the surface of some cells; generally used for locomotion.

Flower The ovary of a plant, plus its associated structures.

Fruit Ripened plant ovary or ovaries.

Gamete Sex cell; egg or sperm.

Gametophyte The haploid fraction of a life cycle.

Gastrula Stage of embryonic development in which germ layers are formed.

Gene hereditary unit on a specific chromosome location; determines a specific trait of an organism.

Gene pool All of the heritable genes in a given population.

Genetic drift Chance change in a gene pool.

Genotype All the genetic traits of an organism, whether or not these are expressed as phenotypes.

Genus Classification group of related species.

Gill slits Characteristic of Chordata: openings in the pharynx.

Glycolysis process of breakdown of glucose into pyruvate.

Golgi apparatus Organelle that packages proteins into cell vacuoles. Also called Golgi body.

Gymnosperms Vascular plants with naked seeds.

Habitat Where a population lives.

Haploid Having one chromosome of each type. The number of chromosomes in gametes.

Hemocoel In an open circulatory system, the circulatory cavity.

Hemoglobin Pigment molecule in red blood cells, which loosely bonds with oxygen and with carbon dioxide.

Herbivore Plant-eating animal.

Heterotrophs Organisms that need external food sources.

Heterozygous Describes genotype with two different alleles of a particular gene.

Homeostasis Tendency of organisms to establish and maintain a balance between internal systems and external environment.

Homologues Chromosomes that pair with one another during prophase I of meiosis.

Homothermy Describes warm-blooded animals, capable of maintaining constant internal body temperature.

Homozygous Describes a genotype with two identical alleles of the same gene.

Hormone Substance secreted directly into the bloodstream, which affects a target organ or organs.

Hydrocarbon Organic compound consisting exclusively of hydrogen and carbon.

Hypothesis A falsifiable statement.

Interphase That part of the cell cycle between one mitosis and the next.

Invertebrate Describes an animal without a backbone.

Leukocyte White blood cell.

Lichen Symbiotic association of an alga and a fungus.

Lipids Fats; also other organic compounds soluble in ether.

Lysosome Organelle-containing protein-digesting enzymes.

Lytic cycle Replication cycle of a virus in which the viruses burst the host cell, releasing thousands of copies of themselves, which infect other cells.

Meiosis Cell division in which the species chromosome number is reduced in half.

Meristem Embryonic plant tissue that continues growing throughout the plant's life.

Mesoderm Middle germ layer.

Messenger RNA (mRNA) RNA used in the cytoplasm for protein synthesis.

Metabolism Organic process through which energy is derived from the breakdown of energy-containing materials (food).

Metamorphosis Dramatic change from a larval stage to adult stage.

Mitochondria Cellular organelles that produce energy.

Mitosis Cellular division that leaves the chromosome number unchanged.

Molecule Made up of atoms; the smallest particle of a compound.

Mutation Permanent change in a gene or chromosome.

Natural selection The weeding out of organisms less capable of survival, leaving those more capable, which live to reproduce.

Neuron Nerve cell.

Neurotransmitters Chemical substance released at the ends of neurons by which nerve impulses are transmitted.

Niche The way of life of a species within a community.

Notochord Structure that defines members of Phylum Chordata; a dorsal nerve cord.

Nucleolus Structure within the cellular nucleus in which RNA is contained and ribosomes are formed.

Nucleotide Phosphate, sugar, and nitrogen compound that is a building block of RNA and DNA.

Nucleus The large cellular structure that contains genes.

Operon A gene or gene group that functions as a unit.

Organelle "Little organ"; a specialized structure within a cell.

Organic compound Any compound containing covalently bonded carbon.

Ovary Female organ that produces gametes.

Oxidative phosphorylation Process by which ATP is synthesized from ADP.

Phagocytosis A process by which a cell membrane engulfs food substances.

Phenotype The visible traits of an organism.

Pheromone A chemical used in communication from organism to organism, typically for purposes of sexual attraction.

Phloem Vascular plant tissue that carries photosynthetic products downward.

Photosynthesis Process by which green plants (and some bacteria) use sunlight to make energy-storing sugars.

Phylum The taxonomic level below kingdom.

Plasmid A piece of DNA that can detach from the main chromosome.

Plastids Organelles in plant cells, including, for example, chloroplasts.

Pollen Structures that enclose male gametophytes in higher plants.

Population The members of a species in a given area that can breed with one another.

Posterior The hind end of an animal with bilateral symmetry.

Primary consumers Animals that eat plants.

Primary producers Photosynthetic plants, which require no external food sources.

Procaryotic Having cells without an organized nucleus or internal membranes.

Protein A fundamental organic molecule consisting of chains of amino acids (polypeptides).

Pseudopods Cellular extensions (as in amoeboid organisms) used in locomotion.

Pyruvate Pyruvic acid; product of glucose breakdown.

Radial symmetry Organismic structural pattern with planes of symmetry arranged about an axis.

Recessive An allele expressed in the phenotype only when two such alleles are present.

Replication Process by which DNA makes copies of itself.

Ribosomes Organelles containing ribosomal RNA (rRNA), which function in protein synthesis.

RNA Ribonucleic acid; a nucleic acid containing ribose sugar, which is essential to protein synthesis and transmits genetic information from the DNA of the cell nucleus to the structures in the cytoplasm.

Saprobe Sometimes called saprophyte. Describes any decomposer organism.

Secondary consumers Animals that eat other animals.

Seed The embryonic sporophyte of a higher plant.

Segregation Separation of dominant and recessive alleles in the offspring of heterozygotes.

Sessile Incapable of motility; attached to the bottom.

Sex-linked trait Genetically carried on the X chromosome.

Somatic cells Nongamete cells.

Species Interbreeding natural populations reproductively isolated from other species.

Sperm The male gamete.

Spontaneous generation The refuted theory that life can spring directly from nonlife.

Sporophyte The diploid fraction of a life cycle.

Symbiosis The mutually beneficial coexistence of two species.

Synapse The structure at the juncture of two neurons.

Taxonomy The branch of biology devoted to organism classification.

Tissue A group of cells related in function and location.

Transcription The copying of DNA information to RNA.

Transfer RNA (tRNA) RNA molecules that arrange amino acids into polypeptide sequences during protein synthesis.

Vacuoles Within a cell, a fluid-filled space surrounded by a membrane; often a storage space for fat or protein.

Vascular plants Plants with vascular xylem and phloem.

Vein In animals, a blood vessel that carries deoxygenated blood back toward the heart.

Vertebrate An animal with a backbone.

Virus RNA or DNA usually surrounded by protein capsid, capable of replication within a host cell.

Warm-blooded Describes animals capable of maintaining a constant body temperature despite the temperature of the external environment.

Xylem In vascular plants, conducts water and dissolved minerals upward from the roots.

Zygote A fertilized (diploid) egg.

CHAPTER 30

RECOMMENDED READING

Arms, Karen. *Biology.* Harcourt Brace Jovanovich, 1997.

Audesirk, Gerald, and Teresa E. Audesirk. *Biology: Life on Earth.* Prentice Hall, 1998.

Brum, Gil, and Larry McKane. *Biology Fundamentals.* Wiley, 1995.

Edwards, Gabrielle I. *Biology: The Easy Way.* Barron's, 1990.

Fried, George H. *Biology.* Schaum's Outline Series, McGraw-Hill, 1990.

Knowlton, David N. *Biology Quick Review.* Cliffs Notes, 1995.

Mader, Sylvia S. *Biology.* McGraw-Hill, 1997.

Solomon, Eldra P. *Biology.* Harcourt Brace Jovanovich, 1997.

INDEX

Note: This index covers Part One: Preparing Yourself and Part Two: Study Guide, pages vii-121 of the text. Material in the sample exams, pages 130-323, is not indexed.